Chris Coyier

# PRACTICAL SVG

## MORE FROM A BOOK APART

**Demystifying Public Speaking**
*Lara Hogan*

**JavaScript for Web Designers**
*Mat Marquis*

**Practical SVG**
*Chris Coyier*

**Design for Real Life**
*Eric Meyer & Sara Wachter-Boettcher*

**Git for Humans**
*David Demaree*

**Going Responsive**
*Karen McGrane*

**Responsive Design: Patterns & Principles**
*Ethan Marcotte*

**Designing for Touch**
*Josh Clark*

**Responsible Responsive Design**
*Scott Jehl*

**You're My Favorite Client**
*Mike Monteiro*

**On Web Typography**
*Jason Santa Maria*

Visit abookapart.com for our full list of titles.

Publisher: Jeffrey Zeldman
Designer: Jason Santa Maria
Executive Director: Katel LeDû
Managing Editor: Tina Lee
Editors: Nicole Fenton, Lisa Maria Martin
Copyeditor: Katel LeDû
Proofreader: Caren Litherland
Compositor: Rob Weychert
Ebook Producer: Ron Bilodeau

ISBN: 978-1-937557-44-7

A Book Apart
New York, New York
http://abookapart.com

10 9 8 7 6 5 4 3 2 1

# TABLE OF CONTENTS

.

# FOREWORD

IN WORKING WITH many design teams and their parent organizations, I've probably seen a hundred different examples of failure and frustration: days of struggling with making a good decision, weeks of wrangling priorities, and months of work lost when an executive objects to the direction at the last minute.

In nearly every case, the root cause of the problem is one of two things (and sometimes both). One has to do with the organization's underlying values, which are incredibly difficult to change. The other results from the team and the stakeholders not having an accurate, shared understanding of the problem they're trying to solve, or of what the attributes of a good solution will be. Thankfully, the latter problem is relatively easy to fix given the right set of tools and an approach that's sensitive to the team's decision-making style.

Building that shared understanding is the purpose of what Dan Brown calls *discovery*.

If you're new to the world of discovery for product definition and design, Dan will introduce you to a wealth of ideas and techniques for making sense—together—of what you need to accomplish. If you're an old hand, these pages will remind you that discovery is a mindset and not just a project phase. Dan's model and the approaches he describes will help you thoroughly frame both the problem and the solution in ways that are—as the title promises—entirely practical.

—**Kim Goodwin**

# INTRODUCTION

NO DESIGN PROJECT starts from scratch. We all come to projects with preexisting knowledge, biases, and assumptions. Even with a little bit of experience, we can feel confident in knowing what works and what doesn't.

But that confidence comes with some uncertainty—an acute awareness that we don't know everything about this particular project, this particular business, this particular audience. Our efforts can't rest on prior experience alone. We can't just dive into creating the final product. We have to start somewhere, building a foundation of understanding and knowledge that clarifies objectives, assumptions, and constraints.

At first glance, that foundation is an expression of the project's goals. As a new project gets underway, you may feel confident you understand the assignment, only to discover you're not sure where you're going. You may find you're not even sure why you're there.

Peel away the layers of the foundation, and it's more than project goals. Design begins not just with a vision of the desired outcome, but also with a statement of how we're trying to help change the user's world, and a characterization of our starting point. The foundation is, in short, an assertion of the problem, a possible solution, and how we plan to get there.

We're responsible for shaping that foundation—no one can hand it to us. Finding that starting point goes by many names. Some people call it *strategy* or *research* or *requirements*. I've heard it called *inception* and *definition* and *ideation*. Whatever you call it, you're learning—learning about users, about the business, about the technology. Learning revs your creative engine: you generate new ideas, you improve upon existing ideas, and you see the problem ever clearer.

I call this *discovery*, because it's as much about the journey as what you find along the way. And, ultimately, it's about uncovering information, and understanding why that information is important. Learning doesn't follow a specific process, and the term *discovery* doesn't imply a particular string of activities. It doesn't imply that some information is more important than other information. What we learn about the target audience is important, but no more or less important than what we learn about technical infrastructure, branding guidelines, or operational constraints.

I wrote this book because I'm fascinated by these early stages of the design process. While writing, I realized something about discovery: it's not a specific process or artifact. It's not a phase or methodology. It's not a school of thought or design framework.

Discovery is an attitude.

This book is about why this attitude is important to design, and how to incorporate it into your work. Whether you're starting a new project or in the middle of one, this book gives you the tools you need to embrace the attitude—so you can define the problem, and start to solve it.

Writing this book helped me articulate three things about discovery that I knew implicitly, but aren't always evident:

- **Discovery frames the problem *and* the solution.** When you define design as problem-solving, you're implying a separation between framing the problem and conceiving a solution. Through that distinction, you're also assuming that discovery is focused only on understanding the problem, the first half of the equation. But I've come to realize that you can't truly understand a problem until you spend some time solving it. You need to try out a few ideas to move forward, and that's okay.
- **Discovery happens throughout the design process.** We are constantly learning. We don't get to a moment and say, "I know everything there is to know, let the designing begin!" We take new ideas and try them out. We have more questions. We mix ideas together. We have even more questions. We discard some assumptions. We test options. We see things in a new way. We validate those perspectives. In design, we're constantly switching mindsets, from confident decision-making to curious knowledge-seeking. We plan projects to suit the business context, but those models don't reflect this meandering path, from learning to deciding and back again.

- **Discovery is a mindset, not a phase.** Oh, it would be nice to compartmentalize learning in a single phase. Time-boxing your efforts to acquire knowledge makes them predictable and cost-effective. But learning doesn't work on a timetable. Your brain forms new mental connections and sees things differently on its own schedule. (Yes, often in the shower.) So, to fit into the modern workplace, we'll concede to having a "discovery phase" or "design sprint"—but the truth is we just don't know when the creative breakthrough will come, or when we'll have to answer more questions.

## Why This Book?

I want you to understand what makes for great discovery, and how it prepares you by providing not just a starting point, but an ending point, too. I won't go into every research technique, every method for unearthing requirements, or every kind of brainstorming activity. There are plenty of books on those things. Instead, I wrote a book to help you tie these activities together, regardless of where or how you work. Design happens in a variety of contexts and scenarios, and it's my responsibility to give you a toolset and vocabulary you can use in yours.

In modern web design and product development, the desire for speed sometimes overwhelms the need for learning. The purpose of this book, then, is to boost your confidence in the decisions you make. Let's start where all complicated things start: a basic definition.

# 1

# DISCOVERY DEFINED

> *Today, the best designs aren't coming from a single designer who somehow produces an amazing solution. The best designs are coming from teams that work together as a unit, marching towards a commonly held vision, and always building a new understanding of the problem.*
> —JARED SPOOL, "The Redesign of the Design Process"

WHEN I STARTED working on the web twenty years ago, no one was talking about the design process. There wasn't a guide to help you communicate design decisions or explore different research activities. The idea of design discovery has emerged over time, taking inspiration and techniques from many other fields. Despite the industry's best efforts, though, it is neither well understood nor well defined.

This is, in part, due to its nature. Discovery has elements of creativity and innovation. We are driven, perhaps, by the myths of innovation, stories of magic and *aha!* moments. But to truly understand discovery, we need to take a more practical view.

Discovery needs to work alongside business processes. We don't have infinite amounts of time or money to explore ideas.

We need to identify the problem, make connections, and deliver actionable insights and solutions as quickly as possible. The tension between the space to learn and explore and the confidence to move forward is what makes discovery interesting.

I define discovery as a set of activities that yield shared knowledge to structure and inform design decisions about a particular project.

## DISCOVERY IS NOT...

I also find it helpful to think about what discovery isn't: strategy, execution, or a single methodology.

### Discovery is not strategy

Strategy is difficult to define, and no doubt many designers will see overlap between my conception of discovery and their understanding of strategy. I see strategy as addressing the problem at a high level without necessarily offering concrete direction for the product or site design. It speaks the language of business, not the language of design. It emphasizes paving a path forward, but doesn't focus so much on understanding the problem—a hallmark of the design process.

Discovery is strategic in that it entails planning and looking forward, but it's grounded in the design process, establishing a vision for your product that's described concretely through architecture and style and tone and layout.

### Discovery is not execution

The more interesting distinction for designers is the part of design that *isn't* discovery, which I'd summarize as fleshing out the details. This is *execution*—elaborating, refining, and implementing the concepts established in discovery. This distinction captures the mindset shift, from learning (about the business and technology and, especially, users) to deciding (about layout and interaction and style, among other things). While execution

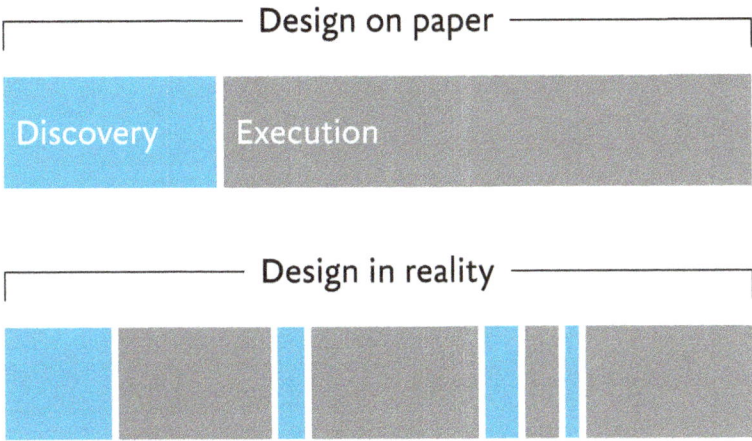

**FIG 1.1:** While it's convenient to think of discovery as the first third of a project, design is more nuanced in practice.

deals with only the latter, making decisions about every nuance, discovery focuses on the former.

It's convenient to think of discovery and execution as two distinct phases in a project (**FIG 1.1**). In this view, the end of discovery is the project's bar mitzvah: a semi-arbitrary point that marks the end of the learning process. But we keep learning as we move into adulthood, and design is much the same.

## Discovery is not a methodology

Over the years, software design and methodologies have changed. Adjustments in approach are generally reactions to earlier failures and perceived demand. We dispose of methods that seem slow and monolithic because we need to "fail fast." Modern design and development practices favor incremental releases, very small units of time, and decreased formality.

What we know about design and creativity, however, is that it thrives when given a chance to percolate. So, we've seen efforts to reconcile the needs of design and modern development practices. We may not be curing polio, modeling DNA, or inventing the printing press, but the problems we're solving are important to someone. And they deserve our best ideas.

Discovery, as I've framed it in this book, is a compromise, a way to squeeze what we know about creativity into what we need in business. It's not a prescribed methodology, but instead a framework for incorporating a learning mindset into your existing processes.

## COMMON DISCOVERY MISTAKES

Discovery is hard, because the first step in any process is hard. Embracing the discovery mindset means putting yourself in a vulnerable position.

You're actively admitting that you don't know enough to solve the problem—that's challenging for anyone. Because you don't know what you don't know, you might focus on unimportant details or pursue the wrong line of questioning in interviews. It's easy to make mistakes, and these are some I've made, time and again.

### Being dogmatic about process

Discovery draws from a range of techniques, organized and prioritized based on a mindset. It should, therefore, fit into any methodology. This book was written to be methodology-agnostic. No tool is perfect for every situation; no framework perfectly aligns with every project.

### Regurgitating requirements

Discovery is more than collecting requirements for all to ignore. Part of discovery is expanding your knowledge: seeking understanding about the context of the project and interpreting requirements in that light. This transforms requirements from a laundry list of things the system should do into a cohesive vision your team can rally around.

## Researching without exploring

Just as part of discovery is understanding requirements in context, another part is exploration: considering different solutions to address those requirements. It's through the tension between trying to understand the problem and trying to solve the problem that the team gets a clear vision of the design direction.

## Thinking of it as "done"

If you work in a typical business environment, you probably have to treat discovery as a phase with an endpoint. Milestones are useful for staffing projects and planning budgets, so you may have to make that concession. That said, discovery isn't ever "done," but then again, neither is your website. You can always learn more about the problem, explore new concepts, or refine your plans.

## Jumping into execution

Even purposeful discovery can seem open-ended or impractical, causing anxiety in some organizations and leading them to skip discovery altogether. It can be easy to rationalize this decision: you know the users, it's a simple product, there aren't any outside stakeholders to manage. Maybe your team feels they have nothing to learn. This is almost never true—and, even if it were, you need discovery to lead the team to a unified vision of what to design.

Finding yourself in these situations can be challenging. The lesson of this book, however, is that discovery is an attitude. Bring that attitude to a meeting, and you can "do discovery" even in the most hostile organization.

## Separating discovery from designers

Discovery entails more than just brainstorming: it's research and planning, analysis and alignment. Some organizations think of these things as outside the scope of design, and structure projects to keep designers hyper-focused on execution (the minute details) or concepts (the big picture without any context). But designers need to be immersed in discovery along with the rest of the product team. No one is dispensable here. The practical argument is time: invest time now and you can avoid backtracking later. But there's a bigger argument, too: you and your team stand to grow from learning through discovery, becoming better designers and collaborators.

## PRACTICE THE MINDSET

Discovery isn't a nice-to-have—it's a necessary part of the design process, regardless of how you incorporate it into your work.

You probably can't control how your business operates, and design problems don't always fit the models described in books, which claim to show you the "right" way of doing design. (This book is no exception, though I try to be less dogmatic).

That said, getting discovery right doesn't entail a specific set of activities or deliverables. If you ignore everything else in this book, perhaps you can rely on these essential elements, regardless of your work environment.

### Build a diverse team

Diverse perspectives help your team explore new ideas, prioritize the right things, and look at the problem in different ways. Invite a variety of stakeholders to participate in the process and include as many people as possible, such as managers, developers, marketers, analysts, sales, and support representatives.

As you analyze and explore ideas, you're bound to uncover different goals and priorities. Healthy conflict is crucial to good design. Working through the details together will strengthen your communication skills and the end product. Discovery gives you space to figure out what you're working on, what your colleagues care about, and how best to collaborate.

## Focus on practical results

Embracing a discovery mindset means putting yourself at risk for navel-gazing or *analysis paralysis:* you might find yourself churning on data you collected, or wanting to do one more interview, or another round of sketches. There's a time and a place for all of this, but there's also a time to get down to brass tacks. The litmus test is at once easy to grasp but difficult to measure: Does your effort help you make better design decisions? The outcomes of your effort serve as tools for the remainder of the design process. If your work isn't leading to better tools, it's time to stop discovery.

## Build a shared language

As designers, we deal with complex problems. Interactive products have lots of moving parts and venture into abstract and philosophical territory. Defining your terms is crucial. You may find, for example, that different members of the team interpret "template" differently. Discovery gives your team a chance to explore ideas and hash out any terminology issues. And team members who participate in the process become more familiar with the problem and develop a sense of ownership as a result.

## Find new approaches

Even if your website's interface ends up relying on tried-and-true design conventions, discovery allows you to feel confident it wasn't for lack of imagination. Through discovery, your team has a chance to experiment, applying new and emerging conventions to common problems, or vice versa. These experiments may feel a little extravagant sometimes, like thought exercises that lead nowhere. But it's important to try them out: you eliminate concepts that don't work, building your confidence that you've chosen the right approach.

### Establish a product definition

Stories of invention and innovation sometimes have a romantic quality to them, characterizing their protagonists as stumbling into a solution. But that rarely happens (if ever). Sure, their work may take detours, and those detours sometimes lead to the breakthrough. They always start, however, with a purpose. We, too, start with a purpose, intent, or objective, and ultimately arrive at a definition of the product, some core essence. This essence, the product's most basic definition, is something that the whole team buys into, that drives execution.

Discovery is, paradoxically, the process by which we determine the product definition. Without a clear purpose or direction, a product is simply a collection of features. An effective discovery process gives your team an ambitious direction—and a shared sense of purpose.

### Encourage consideration

Finally, it's important to note that you can change your mind. Discovery gives you a direction, not a defined path. You can come back to things you've learned, revisit decisions or assumptions, and gather more information as needed. The more you learn, the more confident you'll be. That confidence will also help you deal with conflict and prevent you from taking criticism too personally.

If you can keep these goals in mind, you can do discovery. So it's time to dig into what I mean by "doing discovery."

## DISCOVERY ACTIVITIES

Now that I've defined discovery, explaining its intent and how it's different from strategy and execution, I want to focus on the practical side. There are four types of discovery activities. Depending on where you are with understanding the problem and setting direction, you may need to:

| | FRAMING PROBLEMS | SETTING DIRECTION |
|---|---|---|
| DIVERGENT THINKING | GATHER | EXPLORE |
| CONVERGENT THINKING | PROCESS | FOCUS |

FIG 1.2: Discovery entails four related activities: gathering, processing, exploring, and focusing.

- **gather** new information
- **process** what you've learned
- **explore** different approaches
- **focus** on a particular approach

These categories represent the different kinds of learning you need to do to understand a problem, frame it, and set a direction (FIG 1.2). They're not linear, but interconnected.

Each activity feeds every other activity, moving you and your team toward a clearer understanding of the problem and a more

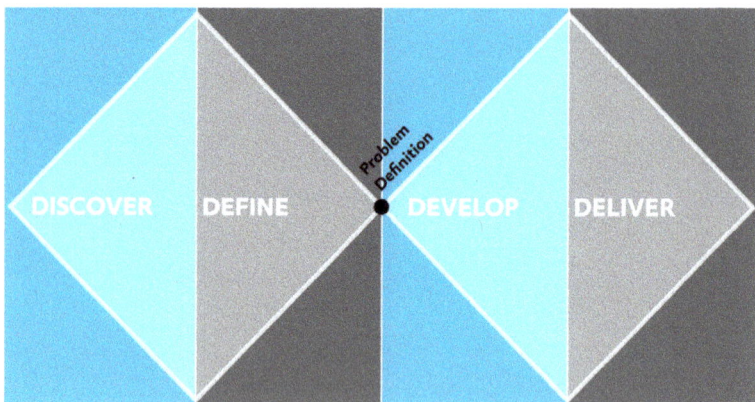

**FIG 1.3:** I first heard about the Double Diamond model from Peter Merholz (peterme.com), though this version borrows language from the UK Design Council (designcouncil.org.uk).

focused product vision. (I think of these quadrants as chambers of the heart, but I say that knowing virtually nothing about cardiology.) Part of the mindset is thinking of every activity contributing to the team's knowledge holistically, not moving a point on a time line—the information you gather may feed your efforts to focus on a definition, and vice versa.

This constant flow of information increases your team's understanding. As you learn something new, you need to decide if it's relevant and how it affects the design. You need all these activities to produce a meaningful starting point for design. And you need the right mindset to embrace this approach to learning.

This is hardly the first time a designer has attempted to depict the creative process. One visualization that has always resonated with me is the *Double Diamond* (**FIG 1.3**), which emphasizes divergence versus convergence. In this view, design entails conceiving of lots of options, then narrowing them down to the one that works best.

Because it positions understanding and solving the problem serially, this model misses some of the nuances of discovery. Some back-and-forth is essential in this process—we solve it a little to understand the problem, and research the problem to devise a more effective solution.

What that back-and-forth looks like for your project (and how much of it you can afford) will depend on:

- how much time you have to gather information and explore ideas
- which activities are appropriate
- how the activities will fit into the project schedule
- what you need to move forward

You'll need to work through these questions with your team, and the coming chapters will help you answer them. So before we get too deep into project plans, let's look at the activities themselves.

# 2
# FRAME THE PROBLEM

> *I guess you have to have a problem if you want to invent a contraption.*
> —THE WHITE STRIPES, "Effect and Cause"

> *By not drawing a clear and compelling problem, you are cheating your team out of an incredible unifying and driving energy.*
> —TOM CHI, IN AN INTERVIEW WITH LUKE WROBLEWSKI

EVEN WHEN I THINK I've seen everything, I learn something new.

I recently worked on a search engine for an industry association. In the Google age, search seems mundane, a solved problem. But my client understood that, while the Google technology used on their site functioned just fine, there were nuances to designing a good search experience for their users.

Before I could recommend a strategy, I needed to understand the problem better. My team interviewed stakeholders, held a brainstorming session, and facilitated user interviews. We gathered information about the organization, about the users, and about what people expected of search. By doing this, we named the demons that plagued the search experience. The demons already existed as complaints and frustrations from employees, association members, and executives. You can't control complaints—but you *can* use them *to* frame a problem.

To summarize the problem, we examined the *challenges* and *constraints* of the project—aspects of the experience that caused the complaints we heard from users. This *exposed a central tension* within the organization that pervaded every aspect of search, and led to a *framework* that inspired and validated our design work.

In discovery, descriptions of the problem space help you frame what you've learned and synthesize it for the rest of the team. To be sure, they also begin to describe the design direction. As I stated in the previous chapter, you're describing both the problem and the solution.

I refer to the outputs of discovery as *assertions*, whether they frame the design problem or describe the design direction.

There are three kinds of problem-framing assertions:

- **Problem statements** tell your team what needs to be fixed. For example, we highlighted that association members were frustrated with the search results because they were out of date and irrelevant.
- **Project objectives** summarize what you hope to accomplish. We wanted to design a search experience that incorporated modern search features and set the stage for personalized results, so the challenges and constraints helped shape the project objectives.
- **Contextual statements** give background information about the users, business, or technology, such as: *Our users are usually looking for information about a job search or industry event. The industry association has several lines of business with different revenue models. The site search uses Google Search Appliance.*

Assertions come in all shapes and sizes. The most important thing to remember is that they aren't set in stone: they reflect your best understanding at the moment. Write them down in a shared file as soon as you have a sense of what they might be, and keep revising them over the course of the project.

# PROBLEM STATEMENTS

Problem statements describe what needs to be fixed, highlighting a pain point, obstacle, or tension that prevents people from accomplishing their goals. They also provide a framework for evaluating ideas, allowing you to ask the question, "Does my concept solve the problem?"

Most problem statements alone are insufficient to evaluate proposed solutions; no single statement can encapsulate all the challenges, constraints, and requirements. But a good problem statement can help you make sure your proposed solution prioritizes the right things.

For the industry association site, our draft problem statement was:

> *Association members are frustrated with the site's search results, because they display out-of-date information and promote irrelevant matches from a narrow slice of the site's content.*

From there, I wanted to understand what was frustrating users, why the information was out of date, and how to include a broader range of content. I continued refining this statement as I learned more.

## Crafting problem statements

Designers struggle to write problem statements because they don't know what will be most valuable to the project team. The key is to focus on the product's users and the challenges they face.

Introduce the project and highlight the core conflict as quickly as possible. I call this "pretending to be a reporter." Summarize the situation for a broad audience—the core team, extended team, stakeholders, and executives—as concisely as possible:

> *The ABC Guild provides up-to-date industry data, job postings, and event information for its members. Members find the Guild's search engine difficult to use and rely on Google to find*

*content. Suffering from poor content strategy and outdated user priorities, the organization's search engine doesn't allow users to search across the range of Guild websites. Since many members are logged in when they use it, they expect the search engine to anticipate their needs, interests, and geographical area. Finally, much of the site's important content is trapped in PDF, making it difficult to index and display in the results.*

Written like a news story, the problem statement reports on a current situation that your team can address. It gives your team something to aspire to: "Our project should fix this."

One way to unearth the underlying problem is through the Five Whys, a widely used technique nicely documented in *Gamestorming* by Dave Gray, Sunni Brown, and James Macanufo (http://bkaprt.com/pdd/02-01/). Start with a simple description of the problem and then ask yourself *why* five times:

- **Problem:** The search engine needs to be fixed.
- **Why?** Members are complaining they can't find anything.
- **Why?** Basic searches yield meaningless or out-of-date results.
- **Why?** Content isn't updated regularly, and the search feature doesn't work consistently across different areas of the site.
- **Why?** Different parts of the organization manage their content in different ways, depending on how important it is to their business.
- **Why?** The organization is so big, different areas of the organization use different revenue models.

Collectively, these "whys" form a problem statement, pointing to the underlying challenge. Do you need to reconcile them into a singular statement? No. By virtue of articulating them in this way, you define the challenges your project is trying to solve. This is what I mean by *assertion*: that a problem statement can take many forms, and may not be a single statement.

You can also use a problem statement template: you fill in the blanks about your target audience, their needs, the product's benefits, key points of differentiation, and perhaps one or two other items. I like keeping it simpler, focusing on the user, their need, and a rationale for the need. Something like:

*[Users] must have [need] to achieve [objective].*

I like being specific with each of these, so *users* isn't just "association members," but "members unfamiliar with the inner workings of the organization."

Another way to craft a problem statement is to position it as a hypothesis, a technique described in the book *Lean UX* by Jeff Gothelf and Josh Seiden (http://bkaprt.com/pdd/02-02/). Using this approach, your assertion is an assumption about the product that you need to test. Crucial aspects of the hypothesis highlight the user need, design implications, and the expected outcome. You're essentially accounting for the design decision, rationale, and expected user behaviors:

*By showing suggested results as users type their search terms, site analytics will show 50% less "pogo-sticking" between search results and site content.*

This is effectively a true-or-false test. Hypotheses build on the basic assertion by proposing a solution, with the purpose of testing it to see if it works.

## Visualizing the problem

Pictures can also reveal the central challenge. I love using pictures because they're instantly engaging. For a current project, I've been documenting the flow of a new business process. Because it's new, much of the detail hasn't been worked out—but my diagram includes call-outs highlighting our outstanding questions (**FIG 2.1**).

Here are a few other ways to incorporate commentary into different kinds of diagrams (**FIG 2.2**):

- **Flowcharts** can show parts of processes that are ineffective, inefficient, or ill-defined.
- **Sitemaps and structural diagrams** can illustrate existing organizations or hierarchies, using them as backdrops for highlighting poorly developed areas of the website.

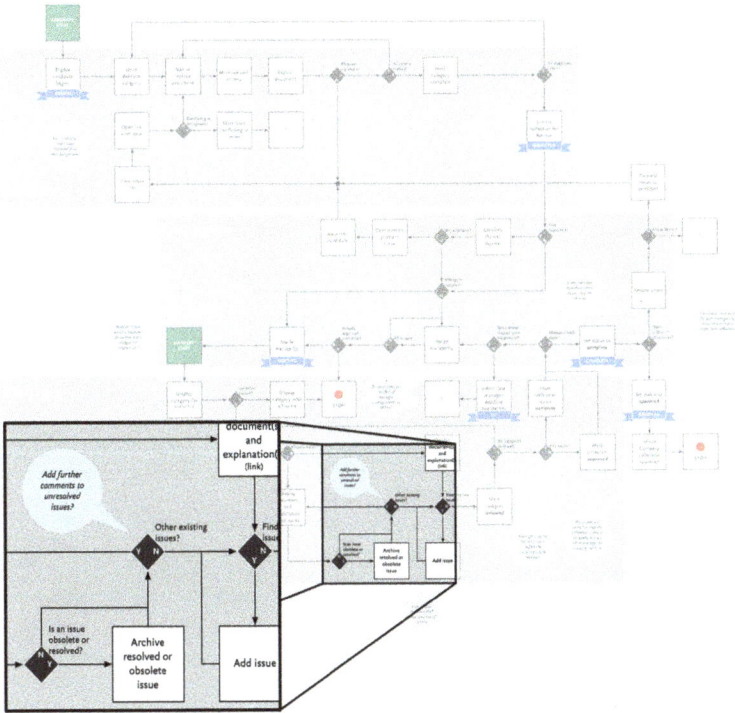

**FIG 2.1:** To distinguish my questions from the actual flow, I use circles with italicized text. Everything else about the diagram is very angular, so the circles imply a separate conversation. Creating this diagram was crucial to our team working through a half-baked business process.

- **Mental models** compare user needs to available content, highlighting the gaps.
- **Concept models** can show the current structure of a business or transaction, and then point out opportunities to support them.

You might call these "problem pictures." Start by drawing a picture of the current state. You might show how ideas relate to each other, how the content is currently structured, or how processes happen. Depict the system as it is today. Then, add

**Flow Chart**

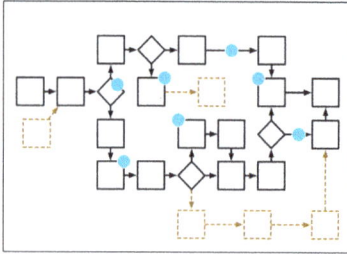

● What's broken    ☐ What's missing

**Sitemap**

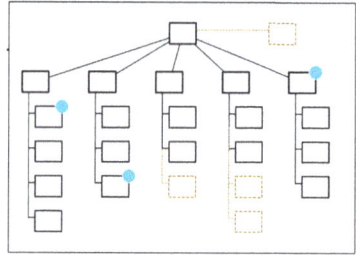

● What's unnecessary    ☐ Opportunities to improve

**Mental Model**

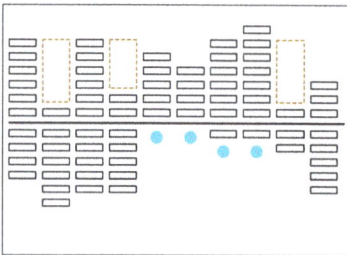

● Missed opportunities    ☐ Insufficient insights

**Concept Model**

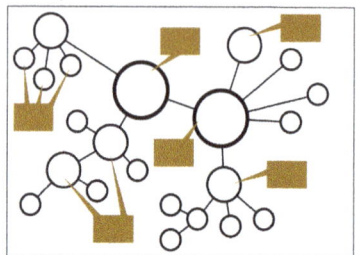

■ How abstract ideas translate to concrete designs

**FIG 2.2:** Diagrams can serve as backdrops to describe challenges and opportunities, placing the design problem in the context of the current state and using color to highlight what's distinct or problematic.

a layer to highlight what's broken, what's missing, or what could be better.

Regardless of how you frame problems, remember that the process is as important as the output. You don't want to spend all of your time wordsmithing—get something down on paper so your team can react to it. Then you'll know whether or not you're aligned.

### Typical problem statement traps

Problem statements generally fall into one of two traps:

- **Too broad:** problem statements may be too broad to be meaningful—for instance, talking about "value to users" without conveying what that means.
- **Too prescriptive:** because it's hard to see a problem without identifying part of the solution, problem statements sometimes venture into the "how" territory.

Look out for these two things, but don't worry about it too much. A problem statement is unifying and motivating, and ultimately malleable. It's a tool to help guide your team, and if it's not working, it's just as productive to dig into why it might not be working.

At the 2013 IA Summit, Stephen Anderson gave a talk called "Stop Doing What You're Told" (http://bkaprt.com/pdd/02-03/), which described mistakes people make when crafting problem statements (**FIG 2.3**). While I think they always boil down to being too broad or too prescriptive, these different types of mistakes offer unique perspectives on problem statements.

## PROJECT OBJECTIVES

Project objectives set the stage for design by aligning expectations around what you'll deliver, not what problem you'll solve. By spelling these things out for the rest of the project team, you can help them understand what matters and what doesn't, like:

*Create a clickable prototype for a responsive web application, using the existing intranet style guide, to help employees track their professional development.*

Project objectives give the team something tangible to reach for. They frame the problem by giving you a sense of what the solution is.

| TRAP | EXAMPLE |
| --- | --- |
| **Anchoring** positions the problem in terms of an existing idea for a solution. | Healthcare providers need a guide to new regulations so they know how it affects their jobs. |
| **Wishlisting** describes the problem as a set of desired features. | Healthcare providers need a guide to regulations, reminders to check in with patients, suggestions on how to deal with particular conditions. |
| **Frankensteining** smushes two or more existing products together to imply a novel solution. | Healthcare providers need Facebook + SalesForce. |
| **Boiling the ocean** suggests that a single problem is really the amalgam of many, many different problems. | Healthcare providers need a shared calendar, to-do list, contact list, and medical encyclopedia. |
| **Amplifying feedback** focuses on a specific message from a non-representative group of users. | Healthcare providers need a guide to Obamacare, because some mentioned they're confused by how it affects their job. |
| **Being presumptuous** makes assumptions about how people will behave, especially with respect to the product. | Nurses will use the software to record every check-in with patients immediately following the check-in. |

**FIG 2.3**: Problem statements can have their own problems. These are the ones I run into time and again. Anderson keeps a complete list on Google Docs (bit.ly/badproblems), where you can add your own.

While you may cringe at the prescriptiveness of statements like these, they can drive conversations about the underlying assumptions, through which you surface the problem. Let's look at the assumptions in this example:

- *Clickable prototype*: build something the team can look at in a browser.

- *Responsive web application*: render the final product in modern HTML and CSS.
- *Intranet style guide*: follow existing design standards, whether they relate to your project or not.
- *Employees*: assume a basic definition of the target audience, but keep an eye out for specifics.
- *Track*: design a system for capturing basic data points about professional development over time, but probe further about the extent of the tracking.
- *Professional development*: don't take it for granted that you know what terms like these mean; every business and industry comes with its own nuances.

By creating a project objective, you express in concrete terms what you and your team need to accomplish. While such statements seem to go against the spirit of design thinking, in a business context where clarity matters, they serve as excellent starting points.

## Crafting project objectives

Ideally, projects defined by objectives like this are about both delivering on the assignment and making sure everyone understands the assignment. Make an effort to put the project into practical terms: methods, milestones, resources, and budget.

Help your team understand success in the context of the project itself, and be specific about:

- **Scope:** Which aspects of the product are you going to focus on?
- **Outputs:** What kinds of deliverables do you need to create?
- **Fidelity:** How closely will the outputs resemble a final product?
- **Time:** How long do you have to spend on each aspect of the project?
- **Priority:** Which aspects should you work on first?

When you write good objectives, you're speaking to the team in terms that are meaningful to their day-to-day work.

It may feel uncomfortable to be so specific at the outset of a project. After all, isn't the purpose of discovery to establish the problem you're going to solve? Shouldn't the design process start with an objective that describes an intended outcome, not a particular output?

On paper, the design process has the luxury of operating this way. But the pressures of reality—time, budget, resources—suggest that we express the output first. That said, project objective statements don't have to be the final recipe for your effort. They serve as a starting point only, guiding your initial activities to clarify the underlying assumptions. Discovery is first and foremost a mindset. Maintaining a skeptical perspective means treating project objectives as hypotheses to be tested rather than instructions to follow.

## CONTEXTUAL STATEMENTS

If problem statements describe needs and project objectives describe a possible end state, contextual statements describe everything else surrounding the product.

Looking at an organization's content management practices, for example, I can declare they use a decentralized process for writing and publishing content. Perhaps this has no bearing on my project for redesigning some of the templates on the site, but it helps me understand the context in which those templates live.

### Business context

You'll start learning about the business as soon as you walk in the door. Broadest in scope, business context assertions describe the systems surrounding, supporting, and drawing value from the product. Through assertions about the business, the design team expresses their understanding of how the product helps the business achieve its goals (make money, save money, attract customers, etc.).

Your team may not need to understand every aspect of the business to do their jobs, but I like getting at least a little bit of information about everything. Even a cursory overview gives me insights into how to make a potential solution fit into the larger context. I don't have an MBA or anything, but these are terms I use to talk about my clients:

- **Domain or industry:** this is the area in which the business operates. Shipping and logistics is a different animal from healthcare. One of my clients deals with licensing architects, which doesn't entail architecture as much as it does understanding state-level bureaucracy. Another client of mine deals with educational content for children—even if my project focuses on the parents' experience, I need domain knowledge about children's publications.
- **Mission:** the organization's mission describes its aspirations in a sentence or two. The mission statement helps you understand the guiding principle that presumably drives every effort at the organization. A Fortune 500 business and my local comic book store may both aspire to dominate their markets, but this means something different to each.
- **Objective:** objectives define achievable goals, usually things that can be measured and reported. They may be big, relating to major changes in how the organization grows or performs, or small, dealing with incremental changes or improvements. Organizations have many objectives, across different aspects of the business, so focus on the ones related to your project.
- **Markets:** markets are groups of possible customers. Markets are different from users: a market has a value and an opportunity, while users have needs and priorities (much more on user context in a bit). Even if your project deals with only one market or segment, it's helpful to have a broad understanding—organizations sometimes distinguish themselves from the competition by looking at markets in a different way. It's also useful to discover the language the business uses to talk about their customers, which may be different from how you want to talk about them. Learning about markets allows you to bridge the gap between your world and theirs.

- **Identity and brand:** often captured in a style guide, the brand describes the way the organization presents itself to its markets, through values and principles, and the mechanics of its identity. If the organization doesn't have a style guide or a brand guide, you can ask stakeholders for a few key words that describe the brand, and then ask how those words relate to your project. Some projects are about developing a new identity or brand strategy. For these projects, capturing the aspirations of the organization is an important part of defining the problem.
- **Culture:** culture is the set of principles, policies, and values that guide how employees behave both inside and outside the organization. It's difficult to learn about this just by asking: you need to see how an organization operates to truly understand its culture. But you don't necessarily need to capture this information formally—simply noting your observations can help you craft documentation and plan meetings more effectively. For example, when working in strongly siloed organizations, you can expect to spend effort and energy encouraging people to share. One product manager was reluctant for me to share conceptual designs, since she knew that once her stakeholders saw an idea, they wouldn't let it go. Understanding this organization's culture helped me work with that constraint, rather than against it.
- **Operating Processes:** finally, the organization achieves its mission and goals through process. Processes entail individual objectives (this is how we publish content to the website) and roles (this is who is involved). Much of your effort in understanding an organization will be in understanding its processes. The way it operates exerts the most pressure on the design process, determining the organization's ability to contribute to the design process and support the product once it's built.

You don't need formal research for all of these things, but you'll want enough background information to understand the constraints the organization imposes on the design process. Getting to know a company as quickly as possible—through domain research, stakeholder interviews, asset reviews, and other gathering activities—is crucial for anticipating how your project will fit into the larger scheme, in terms of both its outcome and successfully getting it done.

Learning about businesses is one of my favorite parts of this job. I feel like an explorer, learning about new corporate cultures. If you get as excited about this stuff as I do, you might find yourself in a veritable black hole of information. Remember: your work is generally in service to creating a website or application, and your project's initial objectives should be your guide for how much digging to do. Keep your research manageable by picking just one or two aspects of the business context to focus on. Be sure to share what you learn with your colleagues: they can shed light or offer new perspectives.

When asserting business context, remember:

- **Don't be the expert.** It can be tempting to learn everything there is to know about the business; that kind of research is fun. But there are people on the team (or adjacent to the team) who know more about how the business operates than you ever will. Lean on them to keep you honest about how things work.
- **Capture the things that impact the design.** Your role isn't corporate ethnographer. (Unless it is, in which case, bear with me.) You're not here to capture every detail, only those things that may influence the design.

## Technology context

Getting to know the business gives us a sense of how a product will create value, but understanding the technology gives us a sense of how it will be built and maintained. You don't need to learn everything about the technology, but you should understand what constraints it will put on your product.

Here are some aspects of the technology you can explore:

- **Data:** for much of our work, data comprises the building blocks. Information, whether tiny fields or large blobs, gives our design life. We can use data to add flavor, but too much and we leave a bad taste; users may feel we've violated their privacy, for example. Using existing systems, constructing new data models, establishing underlying connections between data elements: all these concepts play a role in how I design the system, and also how I can communicate effectively with the team.

· **Content:** Content constraints describe the way content is structured in a system. The existing content management system may prescribe a particular approach for classifying content, or rely on a predefined set of fields that limit how content can be added, stored, and used.

· **Silos:** Organizational silos are often mirrored in technical infrastructure, such that different systems can't talk to one another. Understanding where and how applications are siloed gives you a sense of what challenges you'll face in bridging systems. For example, if your product needs to bring together user data and catalog data, it may be challenging for your product to talk to both databases.

· **Integration:** If you're working on a product that's part of a larger suite of products, you'll want to understand how it fits in. You may need to rely on established flows and user interface elements to maintain a seamless experience.

· **Dependencies:** Working with legacy systems also creates seemingly arbitrary rules about how things work together. It may be that because of how some previous application was built, you need to create a user account before you can let people do anything else, even if that's not the ideal experience for your product.

Technological systems have a direct impact on how people will use the product (and even *whether* they will use it). One perspective your team will take is how to handle technical requirements that put unreasonable expectations on users. Understanding the purpose of these requirements can help you negotiate compromises to the product's design.

As with capturing business context, you don't have to be the expert or capture every nuance. There are a couple other things I do as well:

- **Make the developer part of the design team.** Integrated teams have the benefit of multiple perspectives working together all the time. In addition to keeping you honest about technical capabilities and constraints, a developer on the team can be an enthusiastic contributor to the design process.
- **Don't be afraid to play dumb.** Despite industry pressure on designers to learn even more about technology, multi-disciplinary teams help us to avoid concentrating all our knowledge in one place. Rely on the technologists on your team to explain difficult, esoteric, or proprietary concepts. Encouraging them to answer your questions sets a precedent for information exchange—hopefully they can also come to you for your expertise—whether that's design, audience, business, product, or subject matter, for example.

## User context

It takes a little effort to learn about users, especially if you're seeking firsthand accounts of how they might use the website or product. Finding the right people to talk to and crafting an instrument to extract useful information are activities that take some time. Doing it together, however, means your team builds a shared conception of the target audience.

Different projects call for different aspects of user data, though in building that shared understanding I'm usually after four things:

- **Expectations:** capture what you hear about what users think they'll see or be able to do. On one of my projects, when users couldn't find a particular product in a particular area of a proposed navigation, we noted that the category structure didn't align with their expectations. We used that insight to adjust the categories and do further research.
- **Cognitive constraints:** describe your observations about what's beyond the user's grasp, like imminent distractions, aspects of the UI they didn't see, or information they may not have on hand. When we observe that users "didn't see" a button in a usability test, we're describing a psychological phenomenon: that users were distracted or the button wasn't

in their field of attention. We can use this insight to reduce cognitive hurdles in the visual or interaction design.

- **Scope:** summarize everything users do in connection with the target domain or activity, even if it's not supported by the product you're designing. For example, if we list the tasks that college students do when they research careers, we're articulating scope. We can use this insight to keep the product design focused on the right tasks, or inform conversations about relative priorities of product features.
- **Domain perceptions:** summarize users' understanding of the business or industry, articulating biases or prejudices that might influence their perception of the product. The way people perceive a business or industry can sometimes hold important clues as to their needs, which they may struggle to articulate, or not realize themselves.

Learning about users is both exhilarating and frustrating. For students of human behavior, observing and talking to users—or even gathering quantitative data from them—triggers insights and ideas unlike any other design activity.

Beware the paradox of having too much data and still wanting more. Lots of data may build your confidence in these assertions, but it also leads to spending too much time analyzing the data—analysis paralysis. One of the fatal flaws in our business is that we seem to shortchange analysis, never giving ourselves the time or tools to make sense of what we learn. (That said, the industry has produced a lot of great resources about doing research. Erika Hall's *Just Enough Research* is a great place to start. The Resources list at the back of this book has more [http://bkaprt.com/pdd/02-05/].)

When making assertions about user context, remember:

- **You don't have to use personas.** Personas are summaries of user requirements rendered as caricatures. They were all the rage in the 2000s, but designers came to realize their shortcomings: dehumanizing users, glossing over important details, too much investment in backstory and not enough in design constraints. What matters is that you incorporate users into your thinking, rather than forcing yourself to use a particular deliverable—so choose a format for conveying user data that resonates with your team.

- **It's not just about behaviors.** While understanding how users behave is important, it's also important to understand where, how, and why people are using the product. User journey maps have become a popular way of asserting user context because they evoke the complex relationships people have with products and organizations. If a journey map seems like a big investment, quickly sketch out the major milestones in a user's lifecycle to discuss with your team.
- **User data is a design tool.** Think about how you'll use insights about users as part of ongoing design activities. I use them to kick off brainstorming sessions and as a preface to design review meetings. The format you choose for asserting user context should be one you can use in these and similar situations.

## Content context

There's a fourth context, one that's rarely captured in descriptions of user experience and product design: content.

Content is an aspect of the user's experience that, like interaction and layout and structure and aesthetic, should be designed. That is, every team needs smart people thinking carefully about what a website says and how it says it. (There are now dozens of books on effective writing for the web and establishing a content strategy for your site. I list some favorites in the Resources section.) Content also shapes the design process: when you update a website design, you need to find a way to accommodate all that content.

While you'll be curious about the nature of the content, what it says, and how it's classified, you should try to establish context for the project by focusing on:

- How much content is being preserved?
- How much leeway do we have in changing it?
- How is the content currently stored or structured?
- Which existing content represents extreme cases, in terms of length and age?

- What is the process for making and approving content changes?
- What is the process for creating new content?

The state of the site's content influences the design process. The content creates constraints: boundaries that you have to adhere to (or make a case for breaking) to make the project successful.

Assertions about content generally describe three things:

- **Style, voice, and tone:** How does an organization represent its brand through words? They may have an existing editorial style guide that captures writing standards. This is helpful if you're responsible for creating words, labels, or navigation for the site. Even if the standards have nothing to say about the words you need to create, you can get a sense of what's important to the organization. Perhaps they focus on their name, or translation, or the structure of headlines—clues to what will matter during the design process.

  You may not need to cover editorial style if it's captured effectively in a style guide. That said, you could supplement a missing or inadequate style guide by annotating sample content from the existing site. Including content samples allows you to highlight your findings about the content.
- **Structure and relationships:** As an information architect, I love looking at the structure of content—how the words appear on the site, and whether they follow patterns. Sites with lots of dense prose differ from those that rely on lists. Some sites break up paragraphs using visuals while others maintain the integrity of a complete article, regardless of the story or medium.

  Describe the various formats in which content appears and the intended architecture behind those forms, like the way articles are constructed. Representing these structures can be challenging using just words. I rely on pictures to sketch out different page types. I'll draw maps to show how different page types relate to each other or link together.
- **Scale and quality:** Understanding the amount of content gives you a sense of what you need to accommodate—how

much furniture you've got to move into the house, so to speak. Beyond quantity, though, spending time with the content gives you a sense of its shape and quality. You'll understand how the content fits together, and the kind of story the organization is trying to tell. You'll get a sense of what content gets the most investment, perhaps because there's more of it or because it's the most polished.

Reviewing content is more formally called a content audit; you can use a spreadsheet to express the quality and quantity of what you find. The resulting inventory helps you see patterns and structures, so you can assert how much content is outdated or how inconsistent the tone is.

When you're summarizing the state of the content, remember:

- **Start with questions.** It's easy to think of content context as the output of a content inventory, but you really need to think of it as an answer to questions you have about the content. A content audit is the process of answering the questions, not the process of capturing every piece of content on the site. Your questions should strive to uncover the context—style, structure, and scale—since they exert pressure on the design process. My starting points for this line of questioning include: What content type appears most frequently? How do page structures vary among popular content types? What mental model is implied by the content? How widely used is the style guide?
- **Look beyond the quantitative.** It's easy to focus on concrete assertions like quantity, but content quality impacts design, too. Help your team understand what the content is like and what needs to change, not just how much there is.

## HOW TO GATHER THE RIGHT INFORMATION

I started with assertions—the outcomes of gathering and processing—so you understand what you're striving for as you learn about the project and figure out how to frame the problem. Let's turn to techniques for building these assertions.

Gathering data is perhaps what most people think of as discovery. Gathering is, essentially, research, and research implies the use of specific instruments and data structures. But you may use informal methods, past experiences, or brainstorming activities, which is why I think of this as gathering, not researching.

### Interview stakeholders

*Stakeholder* means "anyone in the organization who cares about the project." (Something about the word stakeholder makes me feel gross every time I use it, but I'm not sure what it is.) By talking to them, you learn what aspects of the project they care about most, what their objectives are, and how the project will affect them. You also learn how the organization works: what happens to data collected, how content gets published, who makes decisions.

Talking to stakeholders can also show you how design fits into the organization. Groups unfamiliar with design, or mistrustful of design, likely require different approaches than those more comfortable with design processes.

Interviewing stakeholders isn't just about what's good for the final product, but what's good for making the project work. There's no special trick to interviewing stakeholders, but like other meetings, you'll get the most out of it if you're well prepared:

- **Embrace your outsider status.** If you're new, an outside consultant, or using the project as an excuse to look at things from a fresh perspective, say that up front. By setting this tone, you create a dynamic where the stakeholder will be patient when you ask them to explain themselves.
- **Respect their time.** Make sure they haven't sat in other interviews like this one recently. Companies sometimes subject the same employees to internal research over and over again. I suspect this happens because some people are considered more forthcoming than others. Before agreeing to interview someone, find out when they were last interviewed (or surveyed) about a similar topic. Also, keep the interview to thirty minutes. That may not seem like enough time, but

a concise interview is better than no interview, and much better than one that meanders. If you've got more material to cover, you can always ask for more time.

- **Ask about their work.** Ask them about the processes in which they play a role, the tools they use, the things they produce or make, the decisions they're responsible for. Ask them about the people they work with and the challenges they face. Ask them what they love about their work and what they hate about it. No detail is too small: ask about the specifics of how work gets done.
- **Ask them to explain.** Take nothing for granted. When they say they're an author or editor, or that they work on the website or answer emails or manage a team, ask them what they mean. Pick any verb or noun they utter and ask, "Tell me more about that." If that makes stakeholders wonder if I'm familiar with the topic, I'll say, "I am, but why don't you tell me about it in your own words."
- **Be a great listener.** Your project entails more than just this one stakeholder, and you likely have to cater to many people, not to mention the product's target audience. But, just for the space of this one interview, pretend there's no one more important in the world. Active listening entails not just being quiet, but prompting, encouraging, and reflecting. Ask them to go on, tell them you're getting great stuff, and demonstrate you heard what they said.

You can interview stakeholders in groups, but I generally like to get them one-on-one, especially if the purpose of the conversation is to gather different perspectives on the project. Some organizations are so political, they might prevent your team from talking to important stakeholders. In these situations, I like to remind the powers that be that anyone we don't talk to now could become an obstacle later. I can negotiate these conversations by offering to interview several people in a group or with a core team member present.

## Conduct secondary and domain research

As an outside consultant, I have to get up to speed quickly on a new client's industry or domain. While this effort hardly constitutes a substantial portion of my discovery time, it's important for me to get to know how businesses in this area work, how they regard their audiences, and how they generally talk about what they do.

Besides conducting my own research, looking at competitive sites and industry publications online, I ask the organization for any research they've conducted independently. Working with people inside an organization, I've learned that this is as much a part of their work as it is mine. They spend time staying up to date on the industry and the competition. They read outside research reports that dig into how the industry markets to their audience.

You're looking for clues about things that will exert pressure on the design process, like constraints that come from how the industry operates. You're also looking for how this particular organization differentiates itself. At the very least, you're learning the jargon and basic concepts so you can have meaningful conversations with the stakeholders.

This kind of background research is crucial, but not the central effort of discovery, and therefore tends to be more informal:

- **Get a reality check.** After some initial research, schedule a meeting with your primary stakeholder to confirm your understanding of the domain and the organization's role within it.
- **Draw pictures.** Sketch models and diagrams to help you get your head around the domain, and ask your team for feedback.
- **Ask about past research.** What was the consequence of the research—and how did it ultimately affect the organization? That will help you understand what worked and what didn't.

## Review existing assets

One way to learn about the business is to look at existing processes and design assets. Guidelines, policies, and internal documentation reveal what's important to the organization, from how it represents itself in public to how it educates new employees. Look for:

- brand identity guidelines
- design standards or pattern libraries
- editorial guidelines or content standards
- examples of pages or screens or content that best embody the brand or editorial voice of the organization
- technical specifications for existing infrastructure
- code libraries
- training materials for new employees

This information can help you understand where your work will fit into the big picture, and may offer insights into organizational challenges. Most importantly, these assets establish constraints. Unless the project entails reworking editorial guidelines or establishing a new brand identity, your work is likely subject to the standards in these documents.

When you're reviewing assets:

- **Compare documented standards to what's implemented.** Existing standards may not be well enforced. It's useful to examine both standards that are enforced and those that aren't, and ask questions about what makes them successful or not.
- **Ask about authority.** There may be multiple, even conflicting, versions of standards. Find out if one is authoritative, and what precipitated the creation of others.
- **Run this activity in parallel with stakeholder interviews or user research.** Assets can trigger interesting conversations. Review corporate assets before your stakeholder interviews to generate questions and avoid covering topics that have already been covered.

## Conduct user research

Planning a research study is one of the first things I do on a new project, but I can't start interviewing users on day one. The background knowledge gleaned from the previous activities—stakeholder conversations, domain research, and asset review—ensures that my research plan and script effectively fill gaps in my knowledge. Although I may ask users about well-worn topics, I choose not to spend my time with them getting a basic education about their domain.

Your initial conversations with the organization will uncover their preconceived ideas about the audience. Those preconceptions establish a foundation for fact-finding activities, provoking questions like: *Is this a reasonable way to model our audience? What are some other distinctions between these audience types?* These questions form the basis of the research plan, and ultimately inform the study script.

Here are a few things I recommend when designing a study:

- **Prepare a script, but don't stick to it perfectly.** I tend to write out everything I want to say, just to make sure I don't miss anything important. I'll highlight key phrases in the script so that when I'm looking at it, I remember to reference those aspects. But I don't stick to the exact wording, and try to let the conversation run where it may.
- **Challenge their assumptions.** Listen for keywords that signal a lack of confidence. Imagine you're asking someone about what they look for in a career, and they respond, "I dunno. Salary?" That suggests there's more to the story, at least because they haven't thought deeply about it. Follow that response with something like, "Well, let's make a list of everything you could look at, then you let me know which is most important."
- **Run a pilot.** I never regret running a sample interview before digging into the real thing, which lets me make sure the script flows smoothly. Whether I use a colleague, stakeholder, or an actual member of the target audience depends on the project and its budget.

One of the best research activities is getting to see the target audience in their natural environment. Observing users while they do the tasks the product will support is incredibly valuable. Real-world observations, however, require a lot of planning. Casually sitting in a Starbucks to observe people ordering coffee is one thing. Being in an emergency room to watch the trauma team in action is another.

I've only done this kind of immersive research (sometimes referred to as ethnographic research) once or twice in my career. Steve Portigal's book *Interviewing Users* is essential reading for any type of user research, but especially for this kind of immersive work (http://bkaprt.com/pdd/02-06/).

## Evaluate competitors

Previous activities—stakeholder interviews, domain research, user research, and corporate assets review—likely revealed competitors and comparative products. These products and services offer similar value to users, and demonstrate the kinds of features and style users expect.

User expectations are crucial to setting the stage for the design process. Competitive products that offer similar features set a tacit standard for how to do things. Successful products don't reinvent the wheel, but instead rely on conventions. At the same time, your design can't offer the exact same thing— you want to improve upon the products.

You won't be able to walk this fine line unless you understand what the competitors are doing. I like using competitive reviews to see how competitors prioritize different kinds of information or features. I like seeing the navigation categories they use, the language established by and for the marketplace.

## Evaluate the current product

With background from stakeholder interviews and corporate assets, your evaluation of the current product can incorporate insights not just about usability, but also about utility and alignment.

As you look at the product, you're determining both how well it addresses user needs and how well it serves the organization's business. You can also look for hidden agendas: what compromises were made in the design for expediency or politics?

When you're looking at the current product, keep a few things in mind:

- **Look for purpose.** As you use the product or browse the site, ask yourself what the purpose of each page is. This can help you surface the most important problems to the user and the business.
- **Use psychological principles to evaluate interfaces.** Interactive design is grounded in psychological principles like attention (what we're focusing on), distraction (what causes us to look away), and modality (what mode we think we're in). Using these as lenses, I can ask hard questions about the design: *Does it direct attention? Does it introduce distractions? Does it expect users to be in a certain frame of mind?* My first exposure to these kinds of principles came when I read Jef Raskin's *The Humane Interface*; more recent resources for learning about psychological principles can be found in the Resources section (http://bkaprt.com/pdd/02-07/).
- **Look for long-discarded design conventions**, like the nested list of folders from the Windows of yore. While these approaches might have worked ten years ago, there's generally a good reason they were abandoned. This isn't to say that modern approaches to design are objectively or comprehensively better; but older techniques can signal that the product's structure is unnecessarily rigid or exposes too much information.

If necessary, I'll create more robust inventories of a site's contents, functionality, or design system to provide the raw materials for a design project. Depending on the project, I look for all the templates or content types or navigation categories in use. This lay-of-the-land activity lets you get your head around the scope of the project, as well as provide a checklist of everything you need to account for. A new site design doesn't do any good if it misses huge swathes of functionality.

These inventories can be painstaking, so here are some techniques I rely on:

- **Define the purpose.** Conducting an informal review of the product design is different from doing a detailed content inventory. Make sure you understand what you're trying to learn from the review before you expend time and energy on it.
- **Capture meaningful data in a spreadsheet.** You can use the same template you used for your content inventory, but since the purpose of this exercise may be different from the last, adjust your spreadsheet to accommodate the different data.
- **Pair the product review with in-house interviews.** People in the organization, even beyond product managers, bring domain knowledge and can inform how you look at the current product. Talking to other business units can shed light on the organizational constraints influencing the product's design.

## Watch people use the product

I recently did a project for a website meant to help high school and college students identify ideal career paths based on a personality assessment. Before diving into design, we conducted usability testing of the existing product. The purpose of this test was, first, to determine what users might not understand about the personality assessment results and the career advice content, and, second, to learn more about what young adults expect from career advice services.

We were surprised to learn that young adults are forgiving of clumsy interfaces—no one struggled with the existing product. They also have high expectations about the depth and complexity of information being presented. Most important, however, conducting this initial usability test gave us a sense of what modern students expect from career advice, and allowed us to prioritize our efforts.

As you conduct usability testing, here are a few things to keep in mind:

- **Put users in the right frame of mind.** In real life, your user's mindset affects how they skim and scan the page. In testing, your questions do the same thing, priming the user to look at things from a certain perspective. It's not productive to ask people "What do you think of the homepage?" right off the bat. Start by asking questions about the overall task—like, "Tell me about how you approached your job search"—to set the stage for later questions and help users get into a realistic mindset.
- **Ask people to tell their stories.** It's easier for people to report on something that actually happened to them, rather than to speculate. Therefore, many of my questions take the form of, "Tell me about the last time you..." This has the added benefit of contributing to the user's frame of mind.
- **Trust your participants.** Leading questions are the bane of a usability script, but they sometimes become unavoidable, especially if you're seeking feedback on a particular part of the site. I use the language, "Don't let me put words in your mouth," and "Keep me honest," to give participants permission to correct me.

## HOW TO PROCESS WHAT YOU'VE LEARNED

*If there is no formal period of time allotted for design synthesis methods, and no formal deliverables associated with these methods, a strong message is sent to the designer: synthesize on your own time, or not at all.*
—JON KOLKO, "Abductive Thinking and Sensemaking"

Discovery is as much about sharing knowledge as it is about gathering it in the first place. Turning raw data into meaningful conclusions requires putting that information into the context of the project, collecting findings that speak to the same message, extracting themes, and prioritizing.

Processing makes information meaningful and accessible. Meaning comes from interpreting the information in the context of the project. Accessibility comes from rendering information in a form that helps others comprehend those insights.

Give yourself enough time to do these activities: it's easy to put all your energy into gathering information, and neglect making sense of it.

## Find the patterns

Of all the information I collect, I'd peg at least ninety percent as qualitative. Quotes from users, complaints from employees, observations from user testing, excerpts from existing assets: these are all bits of information that don't want to be quantified. What's important is the story behind them.

The story, and its moral, may not be apparent until you find the relationships between all the bits of information. This is called *affinity mapping*: putting bits of information together because they relate somehow. An employee describing a step in a business process may correspond to an observation about users missing a button on a screen. Spending time with these inputs uncovers the bigger narrative.

Sometimes the set of data is so big, I can't do it alone. On a recent healthcare project, my team captured nearly four hundred observations from user research on separate index cards. We then split the stack of cards among six team members and sorted them, afterward comparing and contrasting our groupings. This divide-and-conquer approach allowed us to quickly group several hundred observations while coming to a consensus about the underlying patterns.

This isn't the only method, and there's no specific process for sorting and classifying, but I find these approaches useful:

- **Treat each observation as a separate entity**, usually captured as lines in a spreadsheet (**FIG 2.4**). Move them around into meaningful groups. Let go of preconceptions about what the groups are. You might start by putting two observations together, only to realize later that they make better sense in different groups. Try placing two observations next to each

| | Session | Position | Insight | Group |
|---|---|---|---|---|
| 2 | 1 | Force in Favor | Willingness to spend money | Available resources |
| 3 | 1 | Vision | Tagged content only works if the organization is actively tagging | Content organization |
| 4 | 1 | Vision | Centralized content production | Content quality |
| 5 | 2 | Force Against | Corporate thinks of itself as serving stakeholders only, not s | Entrenched processes |
| 6 | 1 | Forces Against | Fear of change | Fear of change |
| 7 | 1 | Forces Against | Not sharing user research | Fear of sharing |
| 8 | 1 | Forces Against | Management incentives not agile (goals set once/year) | Infrastructure discourages adaptation |
| 9 | 1 | Forces Against | Lack of analysis. Someone needs to come in to look for synergies, duplication, waste, and opportunities | Lack of information |
| 10 | 2 | Force Against | No proof of concept, just idea, build and launch | Lack of planning |
| 11 | 2 | Force Against | Lack of product management organization or leadership | Lack of product management |
| 12 | 2 | Force Against | When business units don't treat each other as equal partners | Lack of trust |
| 13 | 2 | Force Against | Silos | Organizational silos |
| 14 | 1 | Forces Against | Division thinking, checklist mindset, loss of quality | Poorly understood priorities |
| 15 | 1 | Force in Favor | Group experience in publishing | Publishing competency |
| 16 | 1 | Vision | Channel for feedback form noncustomer constituents | User engagement |
| 17 | 1 | Vision | Desired change, how to get info: usability testing, interviews | User research |

FIG 2.4: I used a spreadsheet like this one to make sense of observations from a workshop called "Forces of Change." In two different sessions, we captured our desired vision and the forces working against and in favor of that vision. I subsequently grouped them to develop a framework of obstacles and assets.

other and see what they have in common. As the groups emerge, some will feel more right than others.

· **Start with predefined groupings.** You can treat research questions as groups, for example, categorizing your observations by which questions they answer. You might also use your hunches (see next section) as categories, grouping observations into the emergent themes.

· **Think of it as a design exercise.** Take an hour to brainstorm with your team about different models for segmenting users. Evaluate the segmentation models by asking yourselves what would be most useful in the design process.

· **Consider grouping findings by user type.** As these groups emerge, you can perhaps label them as a segment or persona. Teams may come up with half a dozen personas to represent the variety of people using the product. Each persona represents a different set of traits and behavior patterns, tied together by a common perspective. The persona technique was coined by interaction design guru Alan Cooper; authors Tamara Adlin and Steve Mulder each wrote detailed books about the process (see the Resources section).

· **Don't strive for mutually exclusive groups.** Your observations may reveal that users are better understood on a scale. In my work for a career-guidance website, it would have been tempting to segment users by high school and college—but when it came to career aspirations, this dichotomy was almost meaningless. Our research showed that a better segmentation emerged along the lines of career confidence: some students were more confident in their career path than others.

As much as we hope the process is "gather data, analyze for patterns, write themes," reality is messy. As soon as the data starts pouring in, my mind begins to construct themes. Each new interview or observation or usability test forces me to revise the themes: sometimes removing them or shifting them or reprioritizing them. Discussing the findings with the project team yields further insights—different interpretations of the same data.

## Capture hunches

Design research (compared to, say, scientific research) is meant to inspire ideas, provoke new perspectives, and support assertions. One crucial technique is to write down your impressions as they come. Your impressions may not last through the project, changing as you gather more information. They may diminish in importance as you start to better understand the project overall.

After a couple of interviews with high school kids about their career paths, I had several impressions. One was that guidance counselors play a role in helping students get into college, but rarely engage in broader discussions about career. (The truth is that most guidance counselors are too busy dealing with day-to-day issues to pay much attention to tomorrow, but that's a subject for another book.) Another impression was that the students wanted to balance their passion with realistic career choices, looking for something that could pay off student debt while being emotionally satisfying.

I recorded both of these impressions in a shared Google doc, which I called the *project whiteboard*—a bit more structured than raw notes, but still pretty messy. As the interviews progressed, I found increasing evidence for the first impression—neither high school nor college students reported positive experiences with guidance counselors—but little evidence for the second.

A few suggestions for capturing hunches:

- **Work in the cloud.** Create a collaborative document (like a Google doc or Murally board) before you start your gathering activities. As impressions occur, record them in the document with supporting evidence. As a shared document, your impressions can be validated and elaborated by other members of the team. You keep them apprised of your current thoughts, implicitly drawing them into the process.
- **Review hunches with your team and stakeholders** to see what resonates, and to challenge your interpretation of the data.
- **Learn to let go.** As you learn more, some of your hunches will lose relevance or lack sufficient supporting evidence. Don't keep them around—they'll only distract you.

## Create a day-in-the-life story

Processing is more than just collating observations into digestible chunks, like a theme or segment. While those groupings make insights accessible, they paint an incomplete picture, even when taken together. To balance the pithiness of solitary assertions and segments, I often want to show how the product fits into a larger narrative.

For example, on another project involving college students, we sought to design a product that helped them stay on top of their academic obligations. At the time of the project, American college students had a dropout rate of fifty percent, often the result of feeling overwhelmed by too many competing pressures. The product promised to help them set goals and stick to them. For the design team, it was most important to understand the ways in which competing pressures encroached on academic focus. We created a *day-in-the-life story* to describe how students might use the product, especially when their lives became challenging.

Day-in-the-life stories, or vignettes, are meaningful because they evoke concrete versions of themes. They are relatable because they render your user as a character in a story. I use these in place of personas—a story allows me to weave together various insights and convey the context of usage, aspects that are sometimes difficult to capture in personas. It's important to remember both these elements—concreteness and character—when creating stories. (For more on storytelling in design, look no further than Donna Lichaw's *The User Journey* [http://bkaprt.com/pdd/02-08/].)

Here are some techniques I use:

- **Put yourself in the user's shoes.** It's tempting to write these stories from the product's perspective—how it transforms this person's life. A more valuable design tool, however, depends on your being true to the character, the epitome of your user.
- **Include a dramatic moment.** Good story depends on conflict, on the character facing some challenge, and on using the product, hopefully, to overcome it.
- **Write at least two drafts of the story.** Review each draft with stakeholders and your team and make sure they buy into the overall structure, and that you haven't missed any key events.
- **Use your research.** Pull out good quotes or facts from earlier research activities that you can attribute to your protagonist. These details make the story feel real.

## Describe a task or scenario

For the career-guidance service, we thought a lot about the "after the test" scenario—that is, the flow after students take the personality assessment. Assessing personality and suggesting possible directions for careers is a perfectly fine service; but, ultimately, we sought a way for the service to make itself a crucial part of the student's ongoing career search. What does the service look like when they come back? What is their mindset, what is their motivation for coming back, and what functions could support that scenario?

Drawing on what we gathered in our initial research, we imagined different scenarios for using the product regularly. Sometimes called *use cases* or *user stories*, scenarios make insights meaningful because they position them as part of a process that, presumably, directly relates to the product. They aren't complete narratives, like day-in-the-life stories, but they do have a beginning, middle, and end.

Here are a few techniques I use:

- **Think about the user's goals.** The point of a scenario is to explain how users accomplish a specific task. The user may rely on a variety of tools and products, not just the one you're working on, to reach their objective. Ground the scenario description in real data about what users do to complete a task.
- **Outline the process.** Write out the steps in the process, identifying insights from the data gathered that describe each step.
- **Consider exceptions.** Where the data show that different people use different processes, describe how a step in the process may vary or branch off in different circumstances.

## Visualize the problem

For the search project I mentioned at the beginning of this chapter, I knew that a big part of the problem was how the organization oversimplified search. It's easy to dismiss search as "enter term, get results," but there's so much important stuff going on behind those four words, and we needed to make that clear to the organization.

One of my outputs for that project was a poster-sized diagram that relied on the visual language of the organization's domain—chemistry—as a metaphor. This visualized metaphor helped bridge the gap between search and their area of expertise (**FIG 2.5**).

Though this process led to a formal design document, it didn't have to. Visualizing can be an internal tool to help you make sense of things. A mood board, for example, can help you ask the question, "Do I understand the requirements?" In my case, the process of making that diagram helped me understand which aspects of the search experience most affected the organization's operations.

## Experience Principles

**Let the algorithm work.**

**Assert your messages.**

**Prioritize users over politics.**

**Don't try to replace Google.com**

$$Ct + Md + Tm \rightarrow Mt$$

Content ∙ Metadata ∙ Term ∙ Matches

**CONTENT** is everything on the site that's searchable, and that may be of interest to members and other visitors. Content purity-timeliness, accuracy, and format-yields quality results.

**METADATA** is descriptive information attached to content to make it easier to find and recognize. Maintaining metadata requires validating accuracy and consistency.

**SEARCH TERM** is the input supplied by visitors. It is an "unstable element." In that ACS can track which terms are used, but can't easily interpret what users may search for.

**MATCHES** is the collection of content produced by this chemical reaction. It expressed on the web site through **suggestions** and **results**.

**GOOGLE APPLIANCE** catalyzes the reaction, transforming inputs into a set of matches. Tweaking the algorithm can shift the prioritization of matches.

"HEAT" is ACS's insights about visitors and organization. It allows ACS to differentiate their matches from that of external search engines.

## Suggestions Box UI

**Highlight relevant categories**

**Suggest alternate terms**

**Offer hooks to properties**

green chemistry in Education
green chemistry in Jobs
green chemistry in Advocacy

green chemistry principles
green chemistry careers
environmental chemistry
fuel efficiency
greenhouse gas chemistry

ACS Publications
Chemical and Engineering News

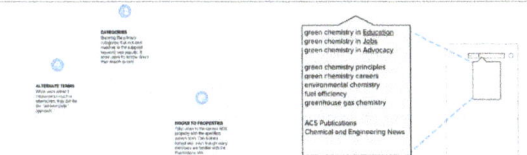

## Search Results UI

**Show powerful results filters**

**Elevate important content**

**Use conventional display for matches**

$$Ct + Md + Tm \rightarrow Mt$$

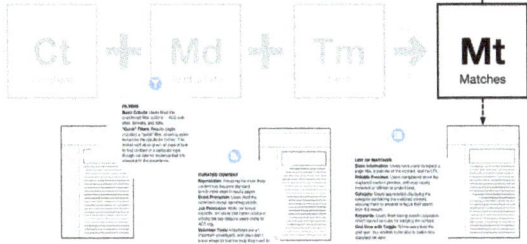

## User Research

**Seek input on content and metadata**

**Understand typical search terms**

**Test display of suggestions and results**

$$Ct + Md + Tm \rightarrow Mt$$

Content ∙ Metadata ∙ Term ∙ Matches

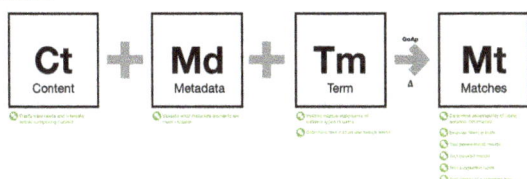

## Content Strategy

**Clean up content**

**Classify and tag content**

**Employ consistent style and structure**

**Archive unnecessary content**

$$Ct + Md + Tm \rightarrow Mt$$

Content ∙ Metadata ∙ Term ∙ Matches

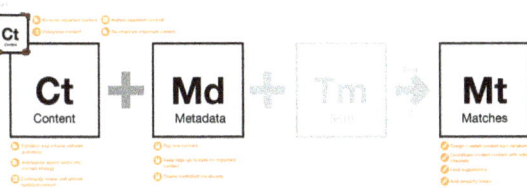

**FIG 2.5:** The search experience rendered as a chemical equation. While the metaphor isn't perfect, it did provide a visual framework for elaborating on different aspects of the search strategy.

The power of visualization isn't just for diagrams:

- Put user research insights on sticky notes to facilitate sorting and classifying.
- Compile a collage of UI examples to compare and contrast approaches to designing a specific feature.
- Use the metaphor of an ecosystem to explore the dependencies and interactions between content elements.
- Sketch a time line to understand the chronology of how users accomplish a particular task or interact with the product.
- Map insights to a two-by-two matrix that lets you relate them on two dimensions. For instance, mapping content to the dimensions "importance to users" and "difficulty to maintain" can help you understand the range of content you're dealing with.
- Create a "bubble diagram" to show the relationships between different concepts, insights, or observations (**FIG 2.6**).

Using visualization means finding ways to make the information accessible through more than words. Diagrams and images are tangible objects that engage your brain differently, offering new ways to surface insights and understand the problem.

Visualizing for understanding is a well-explored topic in design literature, such as Sunni Brown's *Doodle Revolution* and Dan Roam's *Back of the Napkin* (http://bkaprt.com/pdd/02-09/, http://bkaprt.com/pdd/02-10/). (Basically, any book by a Dan or a Brown.) Suffice it to say that pictures are great tools for collaborating and coming to a meeting of the minds. Using sketching as a tool for modeling what you've learned allows you to look at those things from multiple perspectives.

When translating concepts to pictures:

- **Start with a central theme.** Use a singular idea or concept as a visual "backbone" that everything else can relate to—such as my chemistry metaphor for the industry association site's search engine design.
- **Focus on what's essential.** In explaining a high-level user experience, you might be tempted to show how the approach has an impact on the organization, how it draws

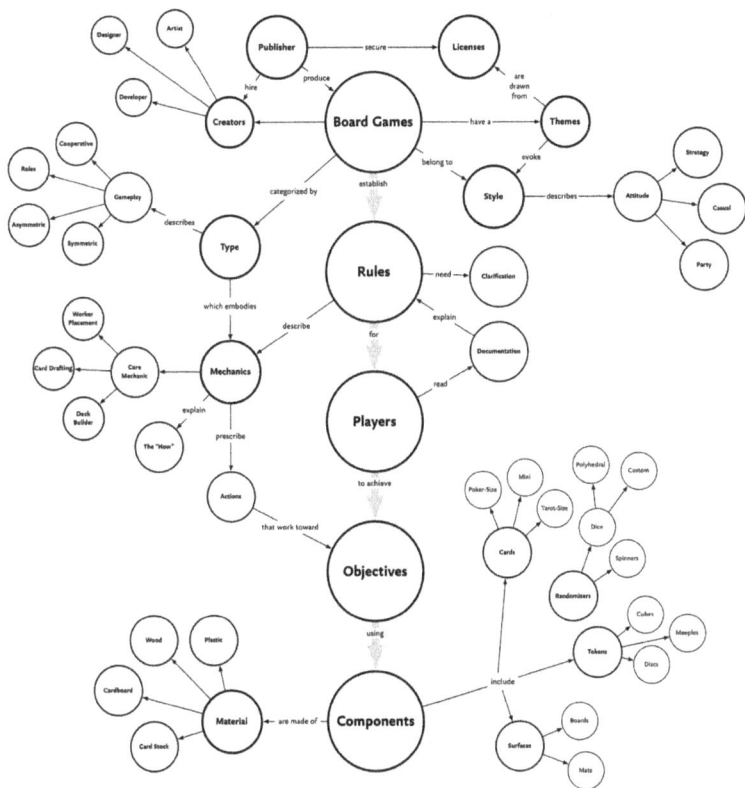

**FIG 2.6:** Bubble diagrams are simple models that let you explore how different concepts relate to each other. By thinking through which relationships to highlight, you gain insights. By focusing on some aspects over others, you look at the underlying system in different ways.

from different data sources, how it meets user needs, and how it improves upon the existing product. It can be tempting to work in *everything* you've learned—and that may satisfy the part of you that likes organizing junk drawers—but, ultimately, too many concepts won't serve to move the project forward.

- **Think about your audience.** In creating the final deliverable for the industry association, I asked my primary point of contact, "What conversations do you need to have?" I was asking how he intended to use the deliverable as a way to facilitate conversations with his customers and stakeholders. Understanding his intended conversations allowed us to shape the deliverable to serve that need.

· **Make it a design tool.** Remember, any deliverable you create as part of discovery is a tool you'll use later for elaborating on the design. As you're visualizing the problem, ask yourself how you intend to use this in subsequent design activities.

## THE PROBLEM WITH PROBLEMS

" *Discovering problems actually requires just as much creativity as discovering solutions.*
—SCOTT BERKUN, The Myths of Innovation

This chapter covered the art and science of framing a problem—making assertions about a project's problem, objectives, and context. There are two key takeaways for framing the problem:

· **Problem-framing artifacts describe what needs improving.** Any artifact you create to frame the problem is really trying to express the context, the objective, or the essential pain point. Personas, sitemaps, and content audits, among others, all play a role in framing the problem.
· **Gathering and processing are not necessarily sequential.** Processing data you've gathered makes sense, but that processing may trigger additional research. That is, even after you process incoming data (through, say, stakeholder interviews), you may need to do further gathering activities (like user research).

Discovery activities entail a back-and-forth, from divergent to convergent thinking, and it's through that process that we fill the gaps in our knowledge. Discovery produces problem-framing assertions, first building up data through gathering, then refining and contextualizing through processing.

|                        | **FRAMING PROBLEMS**                                                                                                                                                    | **SETTING DIRECTION** |
| ---------------------- | ----------------------------------------------------------------------------------------------------------------------------------------------------------------------- | --------------------- |
| **DIVERGENT THINKING** | Interview stakeholders<br>Conduct secondary and domain research<br>Review existing documentation<br>Conduct user research<br>Evaluate competitors<br>Evaluate the current product<br>Watch people use the product | |
|                        | GATHER                                                                                                                                                                 | EXPLORE               |
| **CONVERGENT THINKING** | Find the patterns<br>Capture hunches<br>Create a day-in-the-life narrative<br>Describe a task or scenario<br>Visualize the problem | |
|                        | PROCESS                                                                                                                                                                | FOCUS                 |

**Problem Statements**
**Project Objectives**
**Contextual Statements**

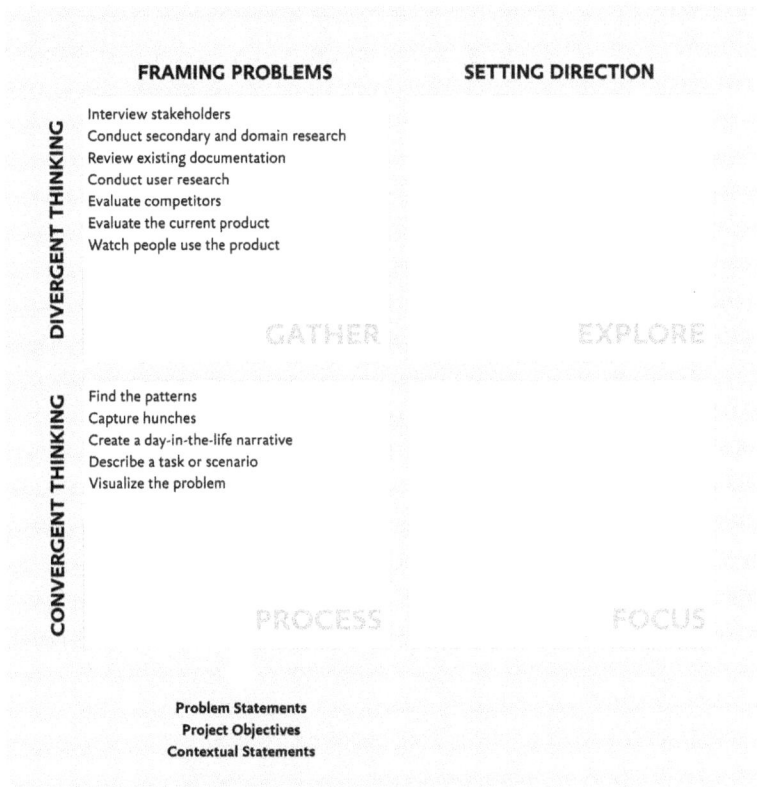

**FIG 2.7:** Framing problems entails gathering information, then processing that information into something meaningful and useful, producing problem statements, project objectives, and contextual statements.

But gathering and processing are only half of the equation (**FIG 2.7**). To truly understand the problem, you have to solve it, at least a little. The next chapter talks about setting a direction, the first steps toward a solution.

# 3
# SET THE
# DIRECTION

> *An essential part of innovation is to envision the new product or service. You have to use it and experience it before it is designed and built.*
>
> —JEFF HAWKINS, FOUNDER OF PALM COMPUTING

ONE OF THE HARDEST design problems I ever worked on was for a company that helps IT groups manage risk. Their product focused on open-source components—inexpensive and widely supported by an enormous community, but often vulnerable to security flaws.

What made this design problem hard was the complexity of the product's underlying structure, a triangle of interrelated abstract concepts. To work through the problem, we created a series of sketches that helped us understand it.

The outcome ended up being a relatively simple prototype, a model of the overall structure of the application. Though we were chartered to create a detailed design, our client later admitted that they knew we wouldn't get there, but that they highly valued our efforts to solve the underlying structure. Those efforts set the direction for everything else on the product.

# DIRECTION-SETTING ASSERTIONS

Much like when we frame problems, we can make assertions that set direction and describe decisions about the design. These decisions will be pretty high-level, meaning they'll deal with a holistic view of the site or product. Decisions about details come later, though you'll see that some assertions get pretty specific as a way of clarifying and testing the direction.

There are three kinds of assertions you can make about design direction:

- **Principles** define what the design should or shouldn't do. These statements are grounded in research, and may be referred to as *implications* when you can tie them to research.
- **Concepts** establish an overall approach for the product, expressed as a central theme or idea.
- **Models** describe the product in an abstract way, showing the underlying architecture, structure, flow, or approach. They offer a sense of how the product will work (without actual functionality).

Principles, concepts, and models represent important decisions, but not especially detailed ones. It's better to make these high-level decisions earlier in the project, because they guide the detailed tactical decisions that follow. For instance, deciding on the tone of the content or the color palette will affect later choices about the words you choose or the colors you apply.

If you try to make tactical decisions too early, you may set a precedent without understanding how it influences what comes next—it's difficult to trace low-level decisions back to a specific objective or problem statement. Why is the button blue? There's no project objective in the world that can justify such a decision.

Instead, you'll make a *few* low-level decisions alongside your assertions, using samples to illustrate, clarify, and demonstrate the application of the high-level decisions. For example, you might arrive at the design principle that the site's tone should be friendly without being too casual or informal. You would demonstrate that through sample screen designs and content, showing messaging that says "Thanks!" instead of the too-formal "Thank you very much" or too-casual "You rock!"

Exploring the big decisions through examples might encourage you to reconsider them, or to find places in the product experience that need variation. Perhaps the color palette is insufficient for everything you need, or the authoritative voice isn't appropriate for certain pages.

By venturing a solution, you're not just asking, "Will this work?" You're also asking, "Do I have enough knowledge to know whether this will work?" That is, steps toward solving the problem may trigger additional insights, or questions, about the problem. Great discovery entails providing just enough shape and definition so the team can get aligned behind them as direction for the product.

## PRINCIPLES AND IMPLICATIONS

Principles are rules that help designers evaluate their decisions about the design. They provide guidance in the form of absolute statements about what the design should or should not do. That said, no set of principles can be exhaustive. They read, sometimes, as commandments: rules that may be applicable to many different kinds of design decisions, and therefore open to interpretation.

There's no industry standard on how to write design principles, so you won't be violating some ordinance if you use pictograms or write a dialogue. But principles are usually just one sentence, often written in the imperative:

*Do more with less (Microsoft Design Principles)*
*Design for the customer and instill confidence (Intuit)*
*Use data to make and improve decisions (Principles for 21st Century Government, Code for America)*

I like these, but they don't feel specific to the product or company. Principles are most powerful when they're directly relevant. These use more elaborate phrases that closely relate to the product:

*More than boxes on a screen (Google Calendar)*
*Transitional interfaces are easier to learn and more pleasant to use (MapBox)*
*Time matters, so build for people on the go (Windows User Experience Design Principles)*

Sometimes, you'll find principles rendered as one- or two-word noun phrases, as if to complete the expression, "The Principle of _____.":

*More Contrast (10 Principles of Codeacademy.com)*
*Consistency (First Principles of Interaction Design, Bruce Tognazzini)*

Principles are sometimes followed by deeper descriptions and examples. My favorite variation of this comes from the Windows User Experience Design Principles (http://bkaprt.com/pdd/03-01/). These principles include questions for designers to ask themselves about design decisions:

*1. Personalization, not customization*
- *Does the feature allow users to express an element of themselves?*
- *Have you made the distinction between personalization and customization?*
- *Does the personalization have to be a new feature, or can it make use of existing features and information (such as the user's location, background picture, or tile)?*

Regardless of the approach you take in framing the principles, use consistent language and structures, if only to make them easier to remember and use. If you lead with a verb, always lead with a verb. If you write a pithy phrase or a complete sentence to express the principle, always do that. If you write single-word principles, well, there's a special place in purgatory for you.

In my practice, I phrase principles as direct consequences of what we learned in research. I call them *implications,* and I prefer them because they fit into the narrative: "We learned

| QUESTION | ANSWER | IMPLICATION |
|---|---|---|
| What are the pain points of the process? | The biggest problem has been that managers get distracted when reviewing an applicant's application. They consider all the additional information supplied by the applicant that's irrelevant. We want them to focus on the work samples. | For managers, show only applicant's work samples and explanations. |
| What parts of the process are you trying to fix? | We often have to return an applicant's application because they provided insufficient explanation for their work samples. | Guide the applicant to provide the right kinds of information for explanations. |
| You often refer to the applicant's work product as samples, exhibits, portfolio pieces. What's the right term? | All these terms work, but applicants and managers seem most familiar with "work product". | Use only the phrase "work product" in the system. |
| How do managers know what to look for in the work product? | We provide guidelines, but they should also work off the applicant's explanation. Managers must confirm that the | Show explanation whenever you show the work product. |

FIG 3.1: Gathering activities generate answers to questions I have about the context or requirements.

that users often lose their place in the system. The implication is that the UI should prioritize clarifying context."

Implications answer the question, "So what?" You've generated a lot of data, and now need to explain why it all matters. I typically document this in a spreadsheet that identifies project questions, answers I've uncovered, and the resulting implications (FIG 3.1).

Ultimately, principles and implications do the same thing, so I won't belabor the distinction between them. In both cases, they make an assertion that, yes, guides the designer, but also provides a test: designers can compare an idea to the principle and determine how closely it adheres to the guide.

There's no standard for design principles, though there are lots of suggestions out there (the Resources section includes a few of the best). Here are my suggestions for crafting design principles.

## Be specific

Principles should be as specific to the product as possible. "Easy to use" isn't a meaningful principle, because it could apply to anything.

For the project with the risk-management company I described at the beginning of this chapter, we used a number of principles. In early versions of their product, users complained that it was easy to lose their place, so they couldn't keep track of what they were working on. This led us to the principle:

*Always display the user's context within the system, so they know where they are and what they're working on.*

Context became something we talked about a lot. It forced us to think carefully before moving information to a different screen, or triggering a dialog box for taking action. Because of this principle, we often asked ourselves, "Can the user tell where they are? Is loss of context here okay?"

## Question your choices

Good principles go beyond specificity: they issue a direct challenge to designers. They force you to take a second look at your work: does the principle invalidate any of your decisions? Done right, principles should make you squirm a little.

In the risk-management product, the complexity of its requirements inevitably produced dense, esoteric designs. Elaborate displays attempted to capture every nuance, pack in every detail. At the same time, our client had heard their users didn't like the dense displays. We had to walk a fine line, and so we relied on this principle:

*Show just enough information to support essential decisions— no more, no less.*

The principle's borderline self-contradiction provoked us to reconsider what stayed on each screen as users worked through the process. *Did we take out too much? Is everything on this screen absolutely necessary?* On one hand, we wanted users to feel confident about where they were, but on the other, we didn't want the page overwhelmed by navigation devices irrelevant to the current task.

We also constantly asked ourselves, "What is 'just enough information?'" and "What are the 'essential decisions?'" Every iteration of the design tested the meaning of these key phrases.

## Inspire your team

Specific and provocative principles may seem like whip-cracking: *Do this, and do it this way.* But a good principle also inspires you, pointing you to even loftier goals. It opens up possibilities by encouraging you to explore—and providing rationale for where you end up.

In Luke Wroblewski's summary of a 2009 talk by Stephan Hoefnagels of Microsoft, he writes, "Goals are the mountain peaks you are trying to get to. [Design] principles are the path we use to get to the top of the mountain."

One of the driving principles for my client's product rested on the insight that the product was focused on bad news: every display was about what was going wrong in the IT department that day, how bad it was, and what wasn't getting done. Like most interactive products, though, this one was meant to be a pleasure to use. In short, we needed to balance the gloom and doom with the satisfaction that comes from understanding the nature and extent of the bad news. We relied on this principle:

> *Build confidence by clearly stating risks and making the data actionable.*

We knew the goal was to help customers manage risk. This principle acted as the path to the top of the mountain by inspiring us to focus not just on reporting the bad news, but also on ensuring customers could do something about it.

## Link principles to research

Principles grounded in research make for stronger statements. The death knell of any principle is arbitrariness: if a principle comes from the subjective preference of the Chief Something Officer or because it reflects the (dysfunctional) way the organization has always worked, designers will ignore it. Your principle can be otherwise perfect, but if its source is suspect, the team won't take it seriously.

The team's participation in all discovery activities is crucial here, too. Since they helped with the research, they can also help with writing the principles. By participating in crafting principles, your team will internalize them. Seeing the principles later will trigger memories of user observations, which they can integrate into their work more readily.

The Windows User Experience Design Principles came directly from research. In reading some of these principles, you can almost hear supporting quotes from users:

- *Reduce concepts to increase confidence*
- *Small things matter, good and bad*
- *Be great at "look" and "do"*
- *Solve distractions, not discoverability*
- *UX before knobs and questions*
- *Personalization, not customization*
- *Value the lifecycle of the experience*
- *Time matters, so build for people on the go*

You might argue that these lack specificity. When you take into account the scope of the project, however—an entire operating system—they're sufficiently provocative and inspirational. "Solve distractions, not discoverability" is a bold statement, offering clear opportunities to refine the design without dictating a particular solution. It opens up conversations, and steers them, too.

## CONCEPTS AND BIG IDEAS

One of my favorite scenes in *Mad Men*, the television show about advertising agencies in the 1960s, is the pitch to Kodak at the end of the first season. Kodak is introducing a new product, a circular tray that makes it easy to store and show photographic slides. They call it "The Wheel," admitting, "We know wheels aren't seen as exciting technology."

Creative director Don Draper, the show's main character, explains that this product isn't about the technology: it's about tapping into our memories and emotions. The agency then pulls the veil off their concept for the campaign: the *carousel*.

By establishing a central concept, a team (whether in advertising or web design) has a singular source of inspiration, a template for considering ideas. And while principles can serve as guideposts, only a concept can establish a vision. With both of them in your toolkit, your team has a potentially interesting tension to draw from.

Using a carousel to describe a slide projector creates a metaphor brimming with meaning and possibility. It shows two ways we can express a big idea:

- **How the product makes you feel:** carousels evoke the joy of reliving happy memories.
- **How the product works:** the spinning carousel mimics storing and displaying photographic slides from a wheel.

Either approach can help us express the big idea behind our digital products and websites. (Though I've never worked on a project that gave us a central concept as elegant as the carousel, which employs both approaches!)

**How the product makes you feel**

The purpose and function of interactive products offer ripe opportunities for metaphors, but metaphor isn't the only way to express a central concept. For one web application project, my team expressed the essence with the phrase, "Power with flexibility." Doesn't quite roll off the tongue like the word carousel, but it evoked the desired feeling: that the app should make users feel like they can do anything.

We elaborated with descriptions of how people would experience unconstrained power with the product:

*Provide users up-to-date status so they feel in control*
*Lower barriers to entry*
*Allow different styles of creating new content*

We also described what "Power with flexibility" meant from the user's perspective:

- **Knowledge:** *having the right data to shed light on immediate needs*
- **Responsiveness:** *being able to provide answers to stakeholders immediately*
- **Accomplishment:** *getting up to speed on a crucial tool right away*
- **Control:** *being able to fine-tune their content to suit different needs in different situations*
- **Comfort:** *seeing the application as an extension of one's own thought process*

Since this essence was a succinct idea, a little elaboration helped it to resonate with both the client and the project team.

## How the product works

Complex interactive products benefit from a central idea that describes how they work. This usually means employing a big idea to convey the underlying structure.

*Shopping cart*, for example, is a popular metaphor used on ecommerce sites. You could use it even if you weren't working on an ecommerce site. The idea of "adding stuff to the cart" is a familiar metaphor that conveys a site's underlying structure. We even relied on this metaphor on our career-guidance site: students would "add careers to their cart" after taking an assessment.

There are a few other tried-and-true frameworks for describing the structure of a website. For web applications, there are two common ones beyond the shopping cart:

- **Hub-and-spoke:** This is perhaps the most common pattern for structuring a website or digital product. The hub-and-spoke metaphor implies that the web application has a central screen, from which users may trigger all other functions.

- **List-detail:** Another typical approach consists of a list of items from which users can select for more detail—like your email inbox.

Do you have to use one of these structures? Of course not. But if your site lends itself to one of these approaches, you have your *big idea* that the rest of the functionality revolves around. (That wasn't a carousel reference. I promise.)

For sites that focus on delivering content (rather than transactional functionality), the tried-and-true frameworks deal more with how the content is organized:

- **topics:** what the content is about, or the subject matter
- **actions:** what tasks the content supports (like researching products versus troubleshooting products)

These aren't the only structures for categorizing content, but they are my go-to starting points.

None of these is a fully fledged design in and of itself. They are well-understood frameworks that serve as the backbone to a much larger design. They are big ideas that describe how the product works.

You don't have to rely on an abstraction or metaphor (like the carousel) to convey the big idea, but instead draw from the emerging library of understood frameworks. That they are becoming part of web design lingo is a testament to their power and flexibility.

## MODELS

Models are any representation of the product without being the product itself. They may rely on the big idea or concept, but they dig into aspects of the site's design. I might use the hub-and-spoke as my central concept, but I use a model to show how that concept gets realized in the scope of my project, with respect to the site's structure, navigation menus, or other aspects of the user interface.

Sketches, diagrams, maps, and prototypes all count as models, varying in their ability to capture and express what the product is, how it behaves, how it looks, and how it feels. They offer different perspectives on a user's experience of the product (or at least the part that you can plan out).

Collectively, models work together to express the solution, or several solutions. Finding the right models—and the right techniques—for explaining the solution is part of the process. For example, wireframes and mock-ups and sketches can create a vision of the final product as shared by an information architect and a visual designer. Models also help your team further their understanding of the problem.

In creating a model, you need to choose which parts of the product to represent. You won't be able to include every screen, error message, or interaction. But, you'll be able to picture the screens, messages, and interactions if the model sufficiently evokes the approach.

The rest of this section describes different types of models, most of which should look familiar (even if you've only just started doing web design). The intent of each model is the same—to assert a part of the user experience. The model should help you make a decision definitively, or assess the implications of that decision. Each type of model serves this purpose a little differently, and the choice of artifact depends on your process, your team, and your need.

## Screen sketches

A lightweight, informal view of a design concept gives you a chance to weigh different ideas and get feedback. Sketches are fast—you can generate lots of ideas quickly—and they can vary in terms of fidelity. That is, you can sketch out a few lines and labels to show the vague idea, or you can be more deliberate about the layout and labeling. With a pair of scissors and an origami mindset, you can show motion or interaction, implying how areas of the screen will swap or change depending on circumstance.

Because they're fast, sketches are great for setting direction. Working with a team, you can all put a bunch of ideas on paper to make some key high-level decisions. That may be sufficient for clarifying the defining aspects of the product.

But while sketches are great tools for quickly sharing ideas, they aren't durable as documentation. The value of a sketch depends on doing it together in the moment. If you sketch independently and share with the team later, the sketch can only take the concept so far. Sketching models the experience very lightly, giving team members a way to focus on a single aspect, like the interface for main navigation.

## Wireframes

Whether wireframes are, at this moment, loved or reviled by the web design community depends on some astrological pattern that remains a mystery to me. Designers can get so worked up about them, but they're an artifact like any other, having their uses and limitations. Wireframes can clarify questions about how users interact with a particular feature. They can show the flow of screens for completing a transaction. They can imply the relative prioritization of content. They can support mock-ups or prototypes to inexpensively describe other parts of the experience.

Like any model, a wireframe can't show the complete experience: you have to deliberately leave out part of the story to focus on the aspects of the solution you want to discuss.

When creating wireframes:

· **Determine the purpose of the wireframes before you start rendering.** Hashing out a flow? Validating content priorities? Elaborating on page-layout variations? Despite the naysayers, there are lots of legitimate reasons to do wireframes, and the most important thing is to know what you want them to accomplish in your process.
· **Make sure stakeholders understand what's being illustrated.** Take a moment before every discussion to remind stakeholders what aspect you're focusing on.
· **Present wireframes alongside visual directions.** Remind project participants that you're pursuing several aspects of the design from different angles.

## Prototypes

When my colleagues and I talk about building prototypes with our clients, we mean an accurate depiction of the website without it being hooked into a content management system, an application, or a database. It's what the website will look like, what it will feel like in the browsers, what it would be like to reach content or accomplish tasks, without building it completely.

I also use the word *prototype* to refer to any series of connected screens, regardless of fidelity. However you define it, people seem to agree that it's more than wireframes—it gives you a sense of how you would move through the website.

Since the web is dynamic, this sense of movement is crucial to understanding the design direction. Product teams use prototypes to reveal this movement, this flow from start to end. One of the deliverables for the project described at the beginning of this chapter was an animated GIF showing how the main interface would work. It was hardly comprehensive, but it allowed everyone to feel what it was like to drill into detailed views.

When you use prototypes, remember to:

- **Define your terms.** With digital product design, *prototype* can be an ambiguous word. Make sure everyone knows what's being delivered.
- **Define your intent.** What do you need to show? How will you use the prototype? Just for demonstrating certain functions? For testing? Make sure you assemble it appropriately for how it will be used in the design process. Highlight the most important features of the product.
- **Treat them like a project.** Prototypes are, in some ways, more than a consequence of the design process: they are small projects in and of themselves, so manage them with rigor.
- **Get the content right.** Prototypes with placeholder content can undermine the effectiveness of the prototype itself. Use real copy—taken from the current site or, preferably, revised for the new one—whenever possible.

## Sitemaps and structural diagrams

Determining the structure of a digital product is among the hardest design decisions to make. Structure has the unfortunate distinction of being both abstract and concrete. In thinking through a structure, designers have to imagine intangible "places" in the product where features and content will "live." In the next breath, they must convey how that structure manifests in the product itself: navigation menus, categories, and contextual linking.

The structure of a digital product is the map of its virtual spaces, the categories in which everything fits. Whether the site is content-driven or transactional, it has a structure. That structure implies what aspects of the site are similar and related. It implies what aspects are most important. And it implies what aspects people will want to use together, a sort of functional adjacency. Working through the structure in a model helps designers get all these things right.

With one foot each in the abstract and concrete worlds, structures are difficult to model. Designers draw all kinds of pictures—like bubble diagrams and sitemaps—to understand and convey a product's underlying structure.

When using structural models, keep in mind:

- **They may be just for you.** The diagram you create to work out the structure may be good for you alone, and won't make sense to anyone not using your brain.
- **They may be thrown away.** Even if you design them to be shared with others, they may not last the project. At some point the structure becomes real as you put the screens together.
- **Keep it focused.** You can't work out every detail of the product in a structural model. Avoid the temptation to solve every single problem in a diagram. Identify the question you want to answer and focus on that.
- **Revel in abstractness.** Structures in digital products are abstract. You'll have ample opportunity to make them real. Enjoy the time drawing circles.

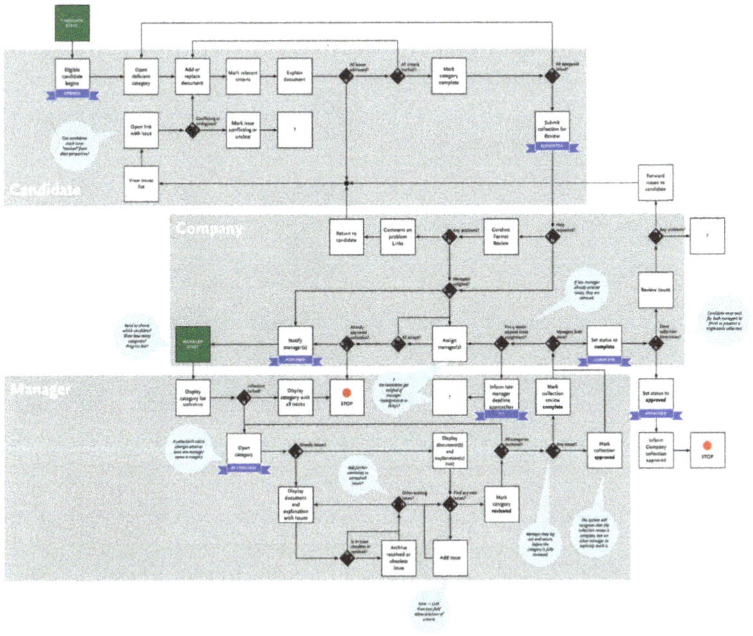

FIG 3.2: Flowcharts conveying the logical progression of steps and choices between end points have been around for decades. In this flow, I'm combining the classic squares and diamonds with additional techniques like swim lanes and callouts to capture the complexity of the application.

## Flowcharts and process diagrams

Digital product design deals with structures in time and space. If structural models describe how concepts and content relate to each other, flowcharts describe dependencies between them: step two follows step one (FIG 3.3).

Though flowcharts aren't reserved for digital products—the previous chapter suggested using flows to document business processes, for example—they effectively convey the big picture of a web application or a function in it. Steps don't necessarily translate to individual screens—models can treat each step in the flow as a distinct screen, or show the subtlety of interaction.

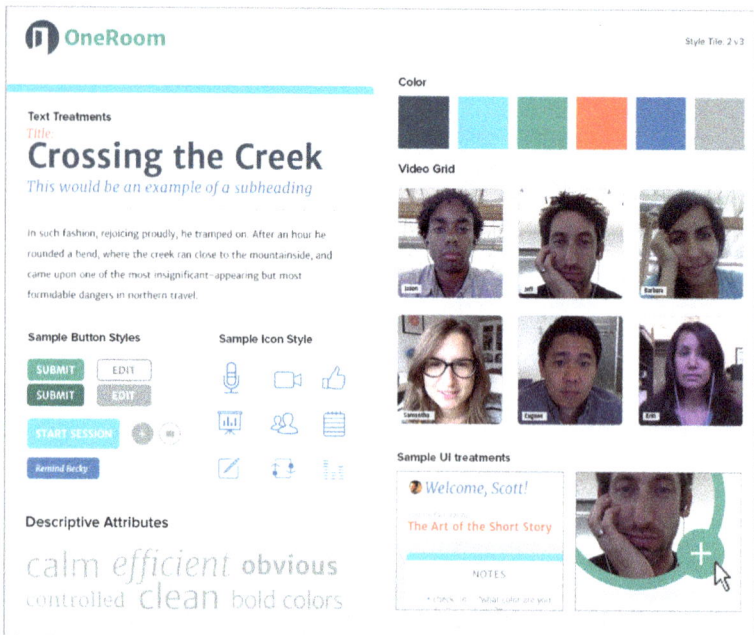

**FIG 3.3:** The style tile is an evolution of the mood board, a way of evoking a particular look through the use of typical web-page elements. At the same time, the tile doesn't serve as the final arbiter of style: those decisions come later. Style tile courtesy of Samantha Warren (samanthatoy.com) and reproduced by permission from JoinOneRoom.com.

When using process diagrams, remember:

- **They may mix the virtual and the real.** Steps in a process can be inside the product or outside the product. These parallels can be worth documenting, but be sure to highlight that distinction.
- **They may not reveal all the consequences.** The real implications of a flow—how long it takes to complete a transaction, dependencies outside the process—may not be evident from the diagram.

## Mood boards and style tiles

As concrete as a product's visual style will be—with color and typography and lines and textures—it starts as abstract as anything else. Visual design is grounded in descriptors, words that convey the brand and character of the product. While such words can be sufficient to drive the visual style of a product, pictures clarify meaning and intention. To convey the visual style, you can assemble artifacts that go beyond words to describe it.

*Mood boards* collect images and other visual elements together in a collage to evoke a style. Designers use them to gauge their shared understanding of the descriptors. When we say "authoritative," do we mean "police state" or "kindly Yoda-like professor"? When we say "warm," do we mean "Hollywood's Tom Hanks" or "the friendly college kids who live next door"?

Since detailed decisions later will depend on these nuances, mood boards attempt to tease them out early in the project.

*Style tiles* (StyleTil.es) are a modern take on mood boards (**FIG 3.4**). Coined by designer Samantha Warren, style tiles perform the same function as mood boards, but relate directly to product design. Unlike mood boards, which are an amalgam of images, style tiles have an inherent structure to represent visual hierarchy. They focus on the central decisions of a visual style: typography, color palette, and visual tone.

Designers have begun experimenting with other techniques to convey early design decisions. Dan Mall uses *element collages* to highlight specific interface elements: why not show a button or a menu or a page transition early in the project? Specific examples set the tone for the design and can be extrapolated to other decisions (**FIG 3.5**).

When using mood boards and style tiles, keep in mind:

- **A little goes a long way.** Avoid weighing down your style tiles or collages with lots of elements. Use just enough to evoke the right tone.
- **Words matter.** Even though this is a visual exercise, providing context with words is helpful. Use words that evoke the aesthetic of the visual direction, aligning the labels and descriptors with the personality.

FIG 3.4: An element collage takes a more granular look at the suggested style, rendering specific portions of the site in a sort of massive web page. This canonical example shows how some elements might transition or animate. Image courtesy of Dan Mall (danielmall.com).

- **Present two or three options.** Presenting just one approach asks stakeholders to say yes or no, and they'll invariably say "no." By presenting options, you encourage stakeholders to talk about which approach they believe comes closer to the project's goals. Even if they offer a subjective opinion, you can bring them back to talking about the options in terms of goals. You risk Frankensteining, but I've found that requests for elements from different approaches leads to good conversations about underlying concepts.
- **Don't have a favorite.** Don't show up with a good approach and a bad approach. You never know what they will pick. Though you can (and should) argue passionately for the approach that seems right to you, you also need to be psychologically prepared for living with the decision they make.

### Sample content

Content is another element of digital products that can go from abstract to concrete very quickly. Early decisions about content characterize the voice and tone of the words and pictures used. The voice and tone may be derived from the same set of brand descriptors as the product's visual style.

The team can prepare sample content to show how that voice and tone is made real—there may be no better way to do so. (Just as teams have visual designers who focus on the visual style, so too may teams have writers or content strategists responsible for providing sample content.) We often prepare sample content in conjunction with mock-ups or prototypes to convey the suggested direction holistically.

With sample content, remember:

- **Focus on the big idea.** Getting the words right is important, but making sure every sentence is perfect isn't relevant for the current exercise.
- **Contrast extreme examples.** You can prepare side-by-side comparisons as you might for visual designs, but some of the best content feedback comes from reactions to extreme examples. When the voice is "whimsical," what's the most whimsical you can make it? Feedback on that helps constrain the direction.
- **Put it in context.** Wherever possible, show the sample content as it might appear in the UI. Seeing the content in context facilitates reactions to it.

## HOW TO EXPLORE POSSIBILITIES

*Liking and not liking can make us blind to what's there. In spite of how we feel about it, it is making its way from the unseen to the visible world, one line after the next, bringing with it a kind of aliveness I live for.*
—LYNDA BARRY, Syllabus

The next chapter will get into how to structure discovery projects. Ultimately, however, it's important to understand these activities as tools you can draw upon as you need them—not toy railroad tracks you click together in a specific order.

Like gathering activities, exploring activities are meant to open up possibilities. While gathering focuses on generating lots of data, exploring activities focus on generating lots of novel ideas. You're turning the insights you gained into possible solutions.

To a certain extent, the activities here are the same as you might find in "creative self-help," a section of the literature meant to "unlock your inner creative." The techniques in this section, however, assume that you're working with a multi-disciplinary team, and that exploring entails engaging with your colleagues.

Exploring will be tempered by focusing activities, tools and techniques for narrowing the ideas into a plausible solution. For now, however, let's look at some of the methods I use to generate lots of ideas.

## Use a common pattern

Web design, in its of existence, has landed on several well-understood starting points for establishing a site structure or a page layout, even for typographic conventions and standardized color palettes. Collectively, these well-known starting points are called *patterns*.

You can express the core concept of the site through a structural pattern, or use patterns to describe the layout of a page and individual UI elements. In the course of your gathering activities, you may discover new patterns, or, more intriguingly, find that competitive sites all follow a particular pattern (which may signal an opportunity to try something completely different).

Dozens of design-pattern books flooded the literature in the last decade, as this idea took hold of the industry. I like Jennifer Tidwell's *Designing Interfaces* best (http://bkaprt.com/pdd/03-02/); flip to any page of that book and you'll find an in-depth discussion of many common patterns, like *wizard* (sequential

| ASPECT OF THE SITE | EXAMPLE PATTERNS |
| --- | --- |
| The structure of the site or application | Hub-and-spoke; Happy path with exceptions; Middle-out hierarchy; Task-driven navigation |
| A single page or screen from the site | Three-column layout; Inbox-style layout; Feed-style layout |
| Visual elements | Google Material button styles; Streamline Icons |

**FIG 3.5:** There are patterns for just about everything in web design. Whether you're talking about the overall site architecture or specific visual elements, there's a pattern for that.

steps for completing a task) or *global navigation* (a consistent set of links to top-level categories).

Try applying common patterns to the raw materials you discovered through processing activities. The solution you create may be far from perfect, but you'll start to get ideas of what might work (**FIG 3.5**).

For the risk-management application I described at the beginning of the chapter, my team and I used an inbox-style interface, with a list of items on the left and details about the selected item on the right. Though the underlying architecture was complex, we landed on a common pattern in part to help simplify the architecture and in part to use a convention we knew users would be familiar with. This concept gave the client a starting point for deeper UI explorations.

When you apply a well-known pattern, remember:

- **Choose any pattern.** Since you're merely trying to stretch boundaries, you're playing a "what if" game, seeing how a particular approach would work for the design problem. Don't censor yourself if you think a pattern will offer useful insights into the problem.

- **Abandon the pattern.** If it's not working, don't try to make it work. This is a thought exercise, a way of seeing various solutions. Don't feel obligated to stick to an approach just because it's canonical.
- **Try two patterns.** Combine two different patterns (what would inbox-style plus hub-and-spoke look like?) to see how they would apply to your problem.
- **Keep a collection of patterns on hand.** I have a stack of index cards on my desk with different patterns I find. I'll flip through these for ideas as I'm working on design problems.

## Relax design constraints

One of my favorite exploration techniques is to remove a constraint, assumption, or parameter of the project. Is one piece of information more important than another? Swap their relative priorities and see what happens. Are we assuming users are not familiar with our content? Assume the opposite and consider how that affects the design. Provocative suggestions like these yield good conversations, forcing you and the team to ask why certain aspects are more important than others.

One common assumption is that two aspects of a product are inextricably linked (known as *pairing*). In working on the career advice application, I was confronted with the pairing that the personality test and career advice were closely tied. I explored concepts that dumped this assumption, treating them instead as two distinct products, each having value for the audience. This freed me from fitting these distinct features into the same user-interface framework. The client found this direction appealing, which in turn allowed them to consider the experience of students who came back to the site to manage their career search. We subsequently explored a unique set of features to support that use case.

When you relax constraints, remember:

- **Imagine the opposite.** If it's difficult to ignore a constraint altogether, consider instead treating it as its opposite.

- **Clarify expectations.** When you use this trick, make sure you tell the rest of the team that you've relaxed a constraint as a thought experiment, to see where it would lead. Project stakeholders may worry you're dumping an entire set of requirements for good.
- **Restore the constraint when you're done.** After you've brainstormed a bit without the constraint, add it back in to see how it affects the ideas you came up with. At the very least, it will spark a conversation about why the constraint is important.

## Sketch together

There are many good reasons to sketch in a group setting, but there are two that consistently stand out for me. First, I get real-time feedback on my ideas. I can use group sketching sessions as a chance to test out my craziest ideas; by introducing concepts that are somewhat outside convention, I can gauge where the team's comfort zone is. That helps me as I focus the design concepts later.

Tied to this is the second reason I love group sketching: I get to peer into other people's minds. People reveal a lot about their perspective and assumptions when they create an off-the-cuff picture. These insights trigger valuable conversations about the product or website itself.

In short, sketching is the perfect way to explore ideas, gather information, and learn more about the problem as you go.

But for those who don't do it very often, sketching can be daunting. To set the group at ease, I emphasize that there are many different ways to share ideas—for example, in previous sessions, I've seen:

- A business analyst who sketched only comparison charts
- A designer who focused on effective labels rather than rough layouts
- An executive who just wrote a list of things
- A team who prided themselves on quantity over quality

Sharing those examples reminds the team that we're not expecting perfectly rendered screens. We're not even expecting well-developed ideas. We're expecting some concepts that give us insights about the problem and possible solutions.

When you sketch in a group, remember:

- **Have a goal.** Before sketching, establish goals for the activity. Figure out what aspect of the product to model.
- **Start with a scenario.** In addition to a goal, it's useful to describe a person using the product—how they're using it, what they want to get out of it, and what their expectations are. By sketching toward a scenario, you establish some reasonable constraints to define the immediate exercise.
- **Set a time limit.** Don't let people draw for more than fifteen minutes. My sketching sessions usually cap the drawing at ten minutes. Like the scenario, this constraint prevents meandering, getting participants to focus on solving the problem.
- Discuss the ideas. The value of sketches is in their speed. Get feedback on the ideas in the sketches before making any final decisions.
- **Extract the good ideas.** Don't assume you'll remember the conclusions about the sketching session from the sketches alone. Make a list of what worked and what didn't.
- **Sketch again.** The cost of iteration is low. Use as many iterations of sketches, with increasing clarity and fidelity, as you need to feel confident about the idea.

## Get some feedback

Feedback is the fuel of creative fire. Without feedback, ideas roar to life, but flame out quickly; with feedback, designers can elaborate, refine, and refactor ideas, constantly adjusting to ensure they address the design problem.

Without structure, though, feedback can spiral into aimless conversations. To focus feedback and avoid counterproductive distractions, follow a few simple rules:

1. **If you're the one giving feedback, don't ask questions.** In other words, after the presenter is done showing the work, you should focus on critiquing the work. Asking questions invites the presenter to defend or explain the design, and takes time away from getting more feedback.
2. **If you're the one presenting work, don't defend your designs.** Once you're done presenting your design concept, keep your mouth shut, and just listen. If you're defending design decisions, you aren't really listening to the feedback.
3. **Limit the time for both presenting and critiquing.** Time limits mean that the presenter needs to focus on the essential parts, and the person giving feedback needs to focus on the aspects of the design that need the most work.

I take these rules seriously; if people ask me questions after my presentation, I'll respond with, "That's good feedback," taking the question itself as critique. If they ask, "What did you mean by this label here?" I won't answer the question, but instead will make a note that the label didn't make sense.

At the conclusion of a feedback session, I ask the presenters, "Do you have enough information to do another iteration?" The point of a feedback session isn't to bruise egos or undermine someone's effort, but instead to give them the information they need to take their ideas to the next level.

When asking for feedback, remember:

- **Be specific about what you need.** When you're presenting something, whether it be a refined design concept, a first draft of a sitemap, or a quick sketch, start with the questions you want answered.
- **Just listen.** Once you've finished presenting the concept, stop talking. The only sounds you should make should be to encourage other people to talk. The most I'll say is, "Say more about that," when someone makes a good point but I need further clarification. Mostly I say, "Good feedback!" or "That's great!" Your job when getting feedback is to record everything that's said, so you can consider it more deeply later.

- **Seriously, don't defend your work.** The best way to shut down productive feedback is to defend your work. I find little more embarrassing in our work than when a designer gets defensive about a napkin sketch. This is especially true in discovery, where the purpose is to open our minds to new knowledge and perspectives. If you feel tempted to defend your work, pause. Listen to what the other person is saying, and ask them to repeat themselves or ask for additional feedback.
- **Teach other people how to give feedback.** Sometimes you'll encounter someone who offers only one-word critiques. In this situation, you need to do more than listen—you need to *actively* listen. Active listening entails using cues to get people to talk more and elaborate. I'll use questions like, "What makes you say that?" or "You said you didn't like this part—can you help me understand why?" Make people feel heard, and then ask for more depth.

## FOCUS ON A SOLUTION

*The key is to look at the viewpoints being offered, in any successful feedback group, as additive, not competitive.*
—ED CATMULL, Creativity, Inc.

Focusing entails not only narrowing the ideas generated through exploring, but also turning them into a concrete plan. The activities in this section are intended to refine concepts into a meaningful direction, which includes a solid definition of the design as well as a plan on how to get there.

### Revise and refine

Exploring entails making a mess—you're composing goals and principles and themes, you're summarizing findings from research, you're sketching screen ideas and navigation concepts. In an exploring mindset, the intent is to grow your stack of ideas.

By contrast, revising is the process by which you choose the best ideas and edit them to even more closely align with the project's goals and constraints. You refine them by transposing them into a format meaningful to the rest of your team. This is a happy place: it's taking raw materials and cleaning them up to express exactly what the project needs.

In the career development project, the user interviews proved to be a gold mine of data. The thirteen young adults I interviewed offered bits of wisdom that produced both broad patterns and interesting divergences to explore. But after accumulating the findings and insights, I edited several drafts of the final report in order to express the results just right. The result was a set of principles we could use during the design process, exposing aspects of career development even the company founders didn't know about.

When revising and refining, keep in mind:

- **Remember your goals.** As you edit, make sure you keep in mind the intent of the output. That is, you're revising and refining for a purpose—to clarify design principles or to describe your project plan. Losing sight of that purpose can lead to bad decisions: you might remove essential details, or elaborate on unnecessary details. Each time you comb through the insights, remind yourself what purpose they'll serve for you, the team, your stakeholders, and the project overall.
- **Revise and refine for the audience.** I don't mean the user of the product you're designing (though that may be the audience for the deliverable). Every deliverable has a target audience—business stakeholders, technologists, other designers—and you're editing to make sure the content resonates with them. (I won't repeat this advice, but it applies throughout this book!)
- **Be careful about perfectionism.** Like any focusing activity, revising and refining can draw you into the quicksand of perfectionism—the desire to wordsmith every sentence, the need to ensure every pixel is in place. Be deliberate about how you spend your time. Don't overthink items that are important but won't see the light of day.

## Prioritize ideas and insights

Exploring generates lots of ideas. Revising and refining gives you one way to focus them, through elimination and consolidation. Prioritizing gives you another way: by elevating the important ones. So the question is, how to prioritize?

Prioritization activities need criteria: the rules you use to say that one thing is more important than another. Since this isn't an exact science, you don't need to draw a straight line between a criterion and what you prioritized—the insights and knowledge you gained in determining the design problem are what you use in this activity.

You could rank features (or screen concepts or design principles or what have you) based on your gut. Taking the time to articulate your criteria, however, can help you facilitate conversations later. Start with an objective: *Given our budget and time line, what features can we build before the end of the calendar year?*

With that goal in mind, you can set criteria, such as:

- strong demand by users, as identified by user research
- close alignment with project goals
- low cost to build and deploy
- supports organization's ability to deliver promotional messaging

Even with solid criteria, your answers won't necessarily pop to the surface. There are three ways you can approach prioritizing, depending on the style of your team:

- **Do this as an individual exercise.** On your own, write down the goals and criteria, then prioritize the requirements, insights, or themes. I may come up with two ways of prioritizing based on slight variations in the criteria, to use as a discussion point with the rest of my team.
- **Facilitate a group workshop.** There are a dozen or so ways to workshop prioritization. One of my favorites, called KJ Technique (http://bkaprt.com/pdd/03-03/), asks participants

to generate a list of ideas, group them, and then vote on the groups. A simpler method, "dot-voting," asks participants to vote for their favorite ideas using colored stickers. These formal methods lead to buy-in because they're structured to make sure everyone's voice is heard.

· **Debate and discuss.** With a smaller group, you can have an unstructured conversation about relative priorities. Articulate and agree upon the prioritization goal and criteria first, but also set a goal for the discussion: "Let's try to get through most of these items" or "Let's get to a provisional prioritization, and we can run remaining disagreements up the flagpole." In these discussions, sometimes it's easiest to eliminate the things at the bottom first, reserving the bulk of your conversation time for talking through any controversial issues. The nice thing about these discussions is that it gives your team a chance to consider the criteria more deeply, surfacing what they mean to each person and why they're important.

When prioritizing, remember:

· **Identify opportunities to gather more information.** When you can't settle on a direction, it may mean you don't have enough information yet. Such gaps aren't a failure of discovery, but a feature. By identifying these obstacles to decision-making, you've found further questions to pursue.

· **Articulate priorities as mantras.** Once you've identified your priorities, make sure you express them in succinct ways to help your team internalize them. No need to break out your inner poet, but short phrases help: "Users prefer completeness over quality" for expressing user needs; "Navigation labels should try to reinforce the brand" for design constraints.

· **Revisit the priorities regularly.** Priorities in one stage of the project may not last. Discussing them regularly encourages your team to keep them in mind, evaluate their ongoing relevance, and ensure they're widely understood.

### Imagine building the site

Discovery isn't just about big ideas—it's about how to turn them into reality. As you formulate a concept, you can plan its implementation. This perspective can help you consider dependencies and level of effort.

Planning implementation has a way of grounding design teams; I call it a focusing activity for a reason. For me, "planning implementation" means roughing out a time line that describes the approach for turning the concept into a fully realized design. Starting with a time line, I have to determine how big the project is and how much effort it will take to design everything. By bringing yourself down to reality, you can keep yourself honest about the ideas you're considering. It can encourage you to seek out new approaches or solutions. It can trigger a need to gather more information, or clarify the design because you don't have enough information to turn it into a real product.

There are three ways you can picture an implementation plan:

- **Start at the final stage and work backwards**, asking yourself at each stage: *What do I need to know to get here?* If I know that my final web application includes a hub screen with six different spokes, as well as a profile and account-management screen, I can imagine the series of steps to get me to that conclusion. I can estimate the level of effort based on unanswered questions.
- **Work forward:** imagine the implementation process as your team gradually building awareness of the product's details. You're thinking about what you will know about the product's design as time progresses. You're asking yourself, *What will I know at the end of one week? At the end of two weeks?*
- **Break an arbitrary length of time into blocks.** Ask yourself: *If I had four chunks of work, how would I break up this project?* In modern parlance, these might be sprints. In the old days, we might have called them phases. Regardless of how you want to structure the chunks, with those chunks in mind,

you have the outline of your story or the overall framework that sets the product's direction.

All three of these approaches give you the same thing: a way of talking about the website in terms of what it takes to make it real.

Regardless of which approach you use, remember what you're trying to get out of imagining the implementation:

- **An awareness of the important chunks.** Taking an implementation perspective forces you to break the product concept up into buildable chunks. Whether you divide things up by feature or functional area or iterative development, you're imposing a different kind of structure on the product.
- **An understanding of dependencies.** Knowing all the pieces of the product, you can also think about the dependencies. Some pieces must come before others. On a recent project, the development team wanted to know when I was going to design the login page. I'd slated this for my last batch because it seemed so innocuous compared to other aspects of the experience. From their perspective, however, everything started with that login experience.
- **An awareness of risks.** Dependencies among product features reveal some risks (*build this before that*). But as you start to picture the execution of the product, you're also asking yourself, *What happens when this goes live? What questions will users have? Where will all the data go? Are there legal ramifications to using the product?* This lens lets you see risks both large and small, and helps you question the foundational direction.

Discovery should yield a plan, describing what it takes to turn the concept into reality. The planning exercise alone gives you a way of clarifying and focusing the direction. By imagining what it takes to make the concept real, you can refine the concept to ensure the direction is feasible (able to be built) and meaningful (contributes to the goals).

## Consider consequences

Sometimes I think of my role as the person who asks "What if...?"

- What if the user searches for an author's name?
- What if a high school student can't find any careers they like?
- What if someone abandons their cart?
- What if the organization creates content that doesn't fit into any of these categories?

It's easy to chalk these up as worst-case scenarios, but Brown's first law of digital product design is: "The exception you haven't anticipated is the exception that breaks your system first."

One way to manage exceptions is to think about the potential consequences of your decisions:

- Decide to exclude a certain type of content from the search engine? Then you need to figure out how to handle searches for it.
- Decide that carts expire after twenty-four hours? Then you need to think through the user behaviors this rule might trigger.
- Decide that kids' content will only be classified by learning objective? Then you need to think about how parents will find books based only on the main character's name.

There's no surefire way to anticipate every exception, and there's no way to think through the implication of every decision. If there were, there would be no buggy software and your college years would be a lot less interesting.

That said, here are some thought exercises I do to try to anticipate implications:

- **Draw it out.** Sometimes we make decisions in the abstract. Representing the decision as a picture—a rough screen design, a partial sitemap, a dialog between user and system—can help you see the consequences of that decision.

- **Narrate your decisions out loud.** Walking through the decisions to people (either inside or, better yet, outside the project) can grant you a new perspective. Explaining how your ideas work, or how you expect people to use the product, often exposes consequences you wouldn't otherwise see.
- **Compare expectations.** Before reviewing your decisions, make a list of what your users expect from the product. Priming yourself with these expectations can help reveal the consequences of your decisions.
- **Assume a role.** Similar to comparing expectations, you can adopt the mindset of someone else in the organization. The purpose here is to think about how other people will be affected by the use of the product: *If I'm a subject-matter expert writing content for the site, what is the impact of this new information architecture on me?*
- **Unearth assumptions.** Every decision comes with assumptions. Exposing and then scrutinizing those assumptions can reveal consequences and implications. Classifying kids' content with learning objectives assumes users will find them meaningful, that there are clear distinctions between them, and that there will be an even distribution of content among them. Thinking through all those things will keep you busy for an afternoon.

## THE PERFECT AND THE GOOD

" *A good idea is a network... If we're going to try to explain the mystery of where ideas come from, we'll have to start by shaking ourselves free of this common misconception: an idea is not a single thing. It is more like a swarm.*
—STEVEN JOHNSON, Where Good Ideas Come From

This chapter covered what makes design fun—coming up with lots of ideas—and what makes design valuable—zeroing in on an appropriate direction, what I call *exploring* and *focusing*. This is the other half of our equation (**FIG 3.6**).

Discovery gives your team a chance to put a stake in the ground with respect to the product's design. You have an oppor-

|                      | FRAMING PROBLEMS | SETTING DIRECTION |
| -------------------- | ---------------- | ----------------- |

**DIVERGENT THINKING**

Use a common pattern
Relax design constraints
Sketch together
Get some feedback

GATHER          EXPLORE

**CONVERGENT THINKING**

Revise and refine
Prioritize ideas and insights
Imagine building the site
Consider consequences

PROCESS          FOCUS

Principles and Implications
Concepts and Big Ideas
Models

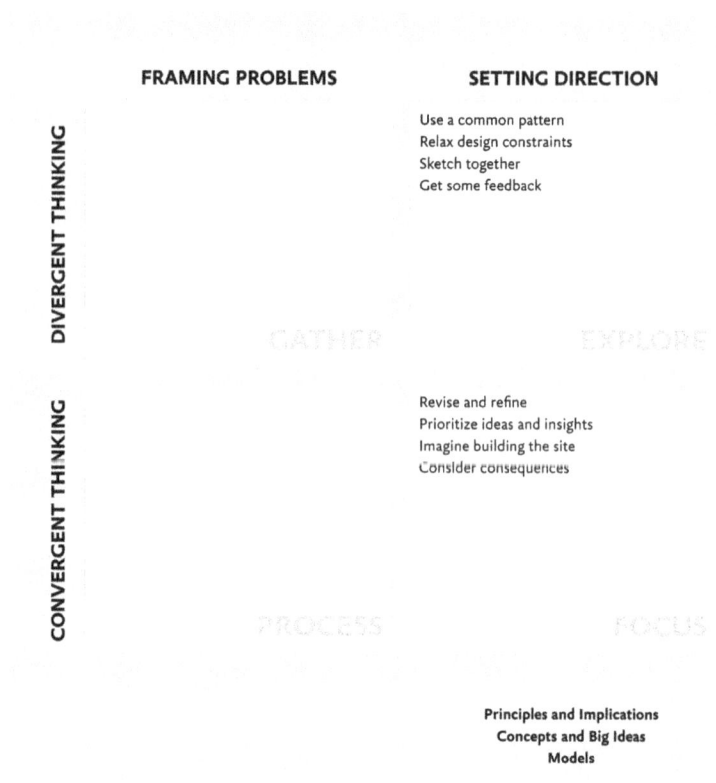

**FIG 3.6**: Setting direction entails exploring and focusing to produce principles, concepts, and models.

tunity to establish a vision that will inform and guide subsequent design activities. It's a big responsibility. And of course you want to get it right.

So it seems that no details are too small—getting them wrong now can have repercussions throughout the rest of the project. At least, it feels that way. But reality intervenes: you'll be reassessing your ideas throughout the project. You'll be reviewing them, critiquing them, massaging them, and refining them to accommodate a requirement that's just emerged, or an insight that's come from the latest evaluation.

Your aim is to specify a direction with just enough information to make sure everyone understands what the product wants to be, what it will look like, and how it will behave. The details matter, of course, but you don't need to spell them all out. And while you're working toward a vision, that vision isn't a goal—it's a direction. Like any journey, the design process can take you a long way from where you started.

# 4 PLAN YOUR APPROACH

> *Every creative person, no matter their field, can draft into service those around them who exhibit the right mixture of intelligence, insight, and grace.*
> —ED CATMULL, Creativity, Inc.

THE PREVIOUS TWO CHAPTERS introduced activities for framing problems and setting direction. Finding the right arrangement for these activities can feel like choreographing a dance while performing it live. You can plan it out ahead of time, but you may need to change course along the way.

If you're a designer, talking about planning may seem a little out of your comfort zone or area of responsibility. Bear with me: design is as much about owning the process as it is about slinging pixels.

## PICKING THE RIGHT ACTIVITIES

When you're planning your discovery work, consider the constraints around time, money, and people. Recall that your objective isn't to have perfect knowledge: it's to have more—and more relevant—knowledge than what you started with.

Time, money, and resources are the most tangible constraints, but as your experience grows, so too does your sensitivity to other kinds of constraints. Here are the ones I deal with most often:

- **Budget:** do what you can afford to do. Discovery represents about a third of your total design budget, but it can be as little as fifteen percent and as much as fifty. Don't be discouraged by small budgets. The constraint forces you to prioritize: *What's the one thing I really need to get out of discovery? What's the one question I need answered to move forward?*
- **Deadlines:** similarly, time is a useful constraint. Discovery, otherwise prone to proceeding without end, benefits from having a clear stopping point. With an established deadline, you have to ask yourself: *What techniques can I use in that time to answer the most important questions?* Driving discovery toward a deadline means you're managing your activities with two metrics: the answer to your question, and the time you have left.
- **Needs:** time and money get you to focus on the right questions; the team's needs get you to focus on the right outputs. How you provide answers to your team depends on how to best communicate with them. Discovery projects of mine have concluded with everything from animated GIFs to InVision prototypes to a series of posters.
- **Resources:** design teams generally have a mix of skills, including research, content, structure, interaction, visual, and technical skills. Your team may be different—some people may wear multiple hats—but it's important to have a good balance and to define clear roles for the project.
- **Leadership:** a team can't be solid without solid leadership. Putting a single person in charge is especially important with discovery. You need not only to coordinate the activities and keep the project going, but also to manage expectations with stakeholders. With so many moving parts, it's tempting to divide and conquer—but since you're trying to set a direc-

tion, a single project lead has a better chance to develop a clear narrative and galvanize the team around it.

- **Politics:** you may want to bring in stakeholders from the extended team for some discussions, but focus on a smaller audience for others. Be careful here and look to your project sponsor or client lead for tips. Involving a broader set of stakeholders is tricky. You need their input, because the project affects them, and you want them to feel confident in the direction. At the same time, more people means more time—and risk. For discrete activities—like brainstorming or reviewing research results—I err on the side of more people in the room. For regular status meetings and discussions about nuts-and-bolts project concerns, I keep it to the core team, where we can strategize about involving more people.
- **Appetite:** a team's interest in discovery depends on their understanding of the design process. If, despite your best efforts, you haven't cultivated that understanding among stakeholders, you may be constrained by their desire to "just design it already." In my experience, even the harshest critics of design process understand that articulating requirements is helpful.

With any of these constraints, reality is your ally. Be clear about what you can accomplish with what you have. If you need to design a social network that reinvents how large corporations share knowledge, you won't have a design direction in two weeks.

## THE BASIC PLAN

The previous chapters talked about the kinds of activities you do during discovery, abstracted to help you understand their purpose. Activities like sketching, getting feedback, and testing familiar patterns all help you more fully understand the problem and begin to set direction. But try to describe a discovery project using those terms and your stakeholders—the people paying for it—may start to tighten the purse strings.

I'll introduce a basic discovery plan, mapping common project tasks to the types of activities we've already looked at, to give you a sense of how discovery activities can start to stitch together. Subsequent sections will talk about how this plan might change to accommodate typical design challenges.

## What every project plan needs

Even the smallest discovery endeavors have critical tasks that should never be skipped. Every activity—gathering, processing, exploring, and focusing—needs to be represented or the whole project is at risk. Every discovery phase should include the following tasks (**FIG 4.1**):

- **Interview stakeholders.** Because every project has people who care about it, interviewing them is the simplest gathering task you can do. These interviews give you a basic sense of requirements and inform other gathering activities as you learn about the business, technology, user, and content contexts.
- **Review materials.** Like interviews, a materials review is a gathering activity you can use to lay down some basic requirements. You might also call this a processing activity because, as you review the materials, you'll need to reconcile conflicting requirements. Review the current product, similar or competitive products, and existing assets like content, pattern libraries, and training materials.
- **Plan and conduct research.** Crafting a research plan requires processing the results from your interviews and materials review. In creating that plan, you're identifying the focus of your user research, the essential questions you need answered. Conducting the research is more gathering, collecting information about the target audience.
- **Analyze data.** In reality, analysis often happens in parallel with your research activities. Process what you learn by finding patterns, capturing hunches, and describing stories and scenarios. Through these tasks, you may articulate a problem statement. You may produce user profiles or some other artifact that makes assertions about context.

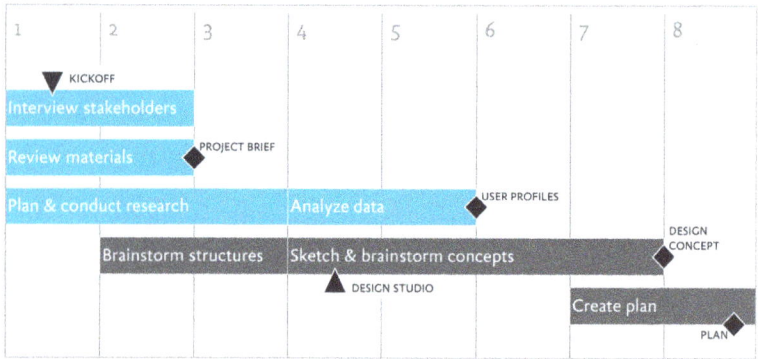

**FIG 4.1:** A basic plan for an eight-week discovery phase.

- **Brainstorm structures.** By working through the application or site's structure, you're at once exploring possible solutions and identifying new lines of inquiry. You might express structure through wireframes, sitemaps, prototypes, or flow-charts, and these visualizations provoke additional questions.
- **Sketch and brainstorm concepts.** Abstract structures are necessary to understand the big picture of the product, but only take you so far. By sketching concepts for screen UI or page layout, you begin to dig into the details, testing some of your ideas and applying some principles. You're a designer, so presumably you can't help but start to visualize the problem. Create mood boards and style tiles; generate sample content; gather feedback; and revise and refine your direction.
- **Create a plan.** Focus your design direction by creating a plan. Describe the final product and a means for getting there, using descriptions that will fire up the organization and be palatable to stakeholders.

The process includes a few crucial milestones, too:

- **Kickoff.** A kickoff is a working session during which the team aligns around the assignment. Beyond working out basic project logistics—like communication, roles, and scheduling—the kickoff is an opportunity to validate the project objective.

- **Project brief.** In most of my discovery projects, this isn't a formal deliverable. Instead, it's a shared cloud document I use to capture the problem statement, objectives, and key insights about context. As the project continues, this shared document might evolve to include the principles, implications, and central concepts.
- **User profiles.** Personas were all the rage for the better part of a decade, and many teams still find them useful. I generally use the more generic term *profiles* to acknowledge that the way I assert user context may not be through traditional personas. Whatever format you use, capture your current understanding of the target audience in terms of their goals and expectations.
- **Design studio.** This is an exploring activity (sketching together) that generates lots of ideas, but also reveals how members of the team think about the problem.
- **Design concept.** At some point during discovery, the team should come to an agreement on the basic principles and central idea driving the design. These may be articulated with both descriptive text as well as design samples, like wireframes, menu designs, or prototypes.
- **Plan.** The final outcome of discovery should be a plan. When finally focusing all the ideas, you should be able to articulate what's next.

## What's missing from this picture

A project plan outlines what activities should happen when, and set stakeholder expectations about when they'll see deliverables. But some of the most crucial aspects of discovery are difficult to fit into a time line. Here are some of those elements:

- **Balancing collaborative and individual efforts.** Discovery benefits from multiple perspectives. Discovery phases should balance group activities with individual activities. Individuals should have easy access to their peers, as necessary to move the design forward. (We call this "frictionless" interaction.) By the same token, individuals should have sufficient heads-down time to process, reflect, and tinker.

- **Critique.** Peer feedback fuels the validation and elaboration of ideas. Designers should cultivate feedback-seeking behavior on everything they do, from product design ideas to planning activities for the team.
- **Project management.** Discovery entails many moving parts, which must be followed and orchestrated. Having a dedicated project manager is crucial. Just because discovery can feel more fluid or open-ended doesn't mean it won't benefit from having a plan. Making sure everyone on the team knows the expected time frame, activities, and outputs can go a long way to coordinate the project. (Even if project manager isn't your title, that doesn't absolve you of responsibility.)
- **Ongoing documentation.** The learning process is more important than the documentation, but the documentation is essential for the rest of the project. Your team can use a shared document to capture impressions, ideas, and insights, a sort of intellectual dumping ground. By lowering expectations of formality and increasing opportunities to share, you create the conditions for new insights: lots of ideas in the same space.

Planning discovery is hard because you don't know when your team will have enough knowledge to feel confident about the problem. The best any of us can do is rely on a project plan and adjust it as needed.

## ADJUSTING THE PLAN

At the beginning of this chapter, I described different factors affecting your approach to discovery, everything from budget to politics. As these factors vary, so too must this basic plan. To explore how to adjust the basic project plan, let's look at five common design challenges:

- **Too much text**: the website must accommodate a lot of content.
- **Poorly defined audience:** the team needs to better understand who the users are and why they would use the site or product.
- **Lack of visual language:** the organization hasn't articulated how their brand extends to their site or product.
- **Unresolved feature list:** the team hasn't prioritized their backlog of ideas.
- **No appetite for discovery**: stakeholders expect designers to design something without properly understanding it.

Though the activities may generally be the same, these scenarios require shifting timing, dependencies, and levels of effort.

## Too much text

Websites with a lot of text include online publications, marketing sites with complex product catalogs, corporate and government websites, educational institutions, and many others. Visitors come to these sites to read. (Social media platforms and blogging applications may be content-heavy, but they focus on content contribution, not just packaging the content itself.)

The scope of a text-heavy site is often measured by how much content it has: the number of pages, the number of templates. To frame a problem for a text-heavy site, your team's activities must include assessing whether the content offered is the content needed by the target audience. You may also be looking at whether the content offered is accessible, through search and navigation, and whether it's maintainable.

The key adjustment to the project plan is accommodating a content audit (**FIG 4.2**); ongoing audit activities anticipate constantly learning and adjusting what you know about the site's content. You'll also need to dedicate time during brainstorming to preparing sample content.

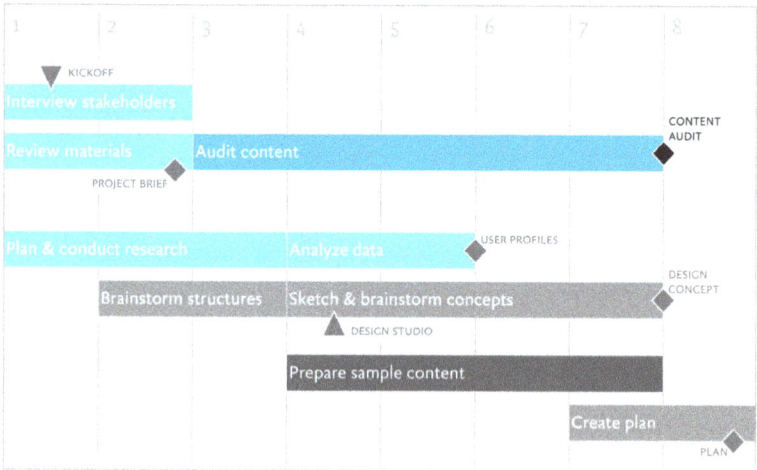

**FIG 4.2:** Adjusting the basic plan to accommodate a content-heavy site.

Larger teams rely on specialists like content strategists to make this determination. Whatever your role on such a project, however, you and everyone on the team need to understand these content-imposed constraints and requirements.

### Poorly defined audience

Occasionally you might start a project already having plenty of insights into the target audience: if you get to work on a popular brand, or a product targeted at the general consumer, or if you've been working inside an organization for a long time. But if you're new to the organization, or working on a different area within the organization, or if you're an outside consultant, you may have no idea what's important to the target audience.

This happened to me when I went to work for a dental insurance company designing an application for their claims processors. (Sadly, my experience didn't give me any more sympathy for the insurance industry.) It also happened to me when I was working on the career advice application for high school and college students—though I'd gotten some career

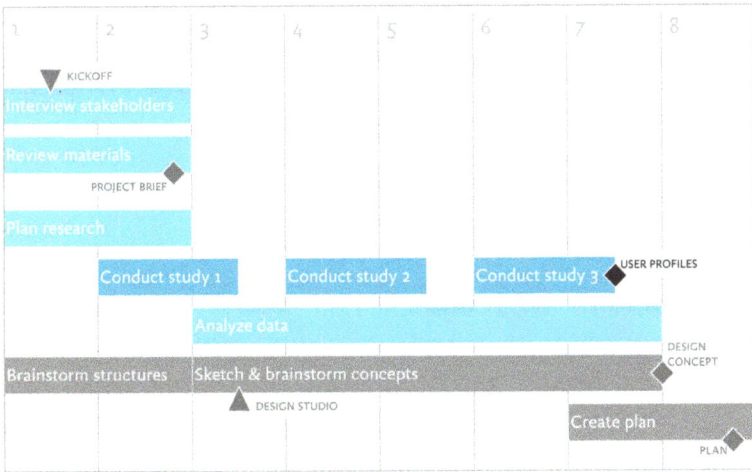

**FIG 4.3:** Adjusting the plan to generate more insights about users.

counseling as a young man, the landscape has changed so much in fifteen...er...twenty-five years.

When the target audience isn't well understood, I like to do multiple research endeavors, what I call *studies* on the project plan (**FIG 4.3**). I might conduct each study as a different research activity—review, survey, baseline usability testing—or repeat the same activity with an increasingly refined script. If I recruit twelve people for research, for example, I might interview four in the first study, then use that data to refine my questions for the next four, and on.

I'll also use different research activities when I want to gather different kinds of information from the same target audience. On a recent project, we did both a card sort and a tree test (testing navigation concepts) to reveal different insights about our users.

The other change to the basic project plan is to move up the exploring activities. By experimenting with structures earlier in the process, you can identify essential lines of questioning for your research. On another recent project, I started diagramming an application flowchart practically on day one to help me get my head around the product.

Those explorations continue, constantly fed by the ongoing user-research studies. As you learn new things, you integrate (or at least consider) those insights into the design concepts. Likewise, further concepts trigger additional lines of questioning for research.

## Lack of visual language

Visual language refers to the way a site design expresses an organization's identity, including color palette, typography, and ornamentation. That visual language flows through every aspect of the web design, from the layout grid and vertical rhythm to the photography and imagery.

When working on a product from scratch, or on a product that needs dramatic improvements to the visual language, discovery work establishes basic visual decisions. You can adjust the project plan to incorporate time for visual explorations, like building style tiles and an element collage (FIG 4.4). My team uses these tools to first establish a high-level style direction, then experiments with applying that style to specific aspects of the UI. Visual designers work closely with UI designers (sometimes they're the same person) to make sure these explorations happen in lockstep.

It bears repeating here that these basic visual decisions may be codified in artifacts like mood boards and style tiles, but it's helpful to create samples that show these decisions in action. All members of the team will want to try this initial direction on for size, especially if brand identity encompasses not just visual style, but voice and tone as well.

## Unresolved feature list

Many projects start with project stakeholders arriving to a kickoff with their wish list of desired features. Discovery is crucial in these projects to prioritize and align those features, and to make sure that selected features hang together in a meaningful way.

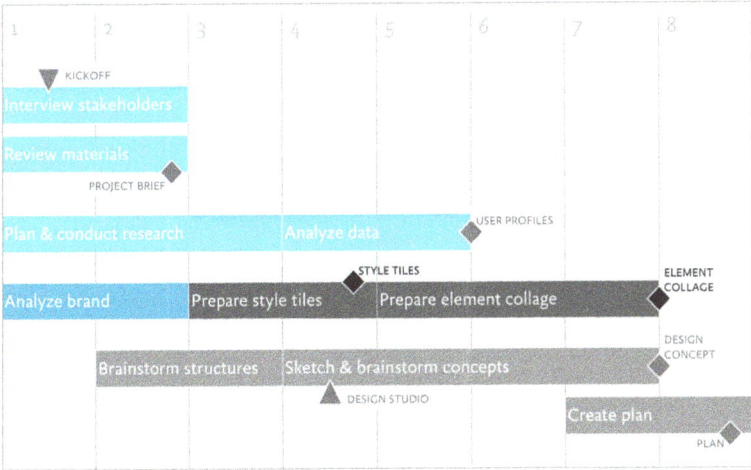

**FIG 4.4:** A project plan with more visual design activities to clarify the requirements and direction for the visual language.

How to get these features prioritized? There will undoubtedly be factors beyond your control, like budget, executive expectations, and marketing pressures. But initial research to understand user needs, form hypotheses, and test those hypotheses can help you influence priorities (**FIG 4.5**).

In my own work, I like to interview users about their pain points, and usability test the current product, to shed light on what features or improvements might be most important. Taking that information, I prototype or mock up the new ideas, then test those in front of real users. Seeing those features or improvements in reality may change users' perspectives or further illuminate their needs.

## No appetite for discovery

In cultures hostile to the discovery process, stakeholders believe that designers demonstrate value only by producing pictures. Leading projects with brainstorming sessions—like the design studio—can help win over such stakeholders (**FIG 4.6**).

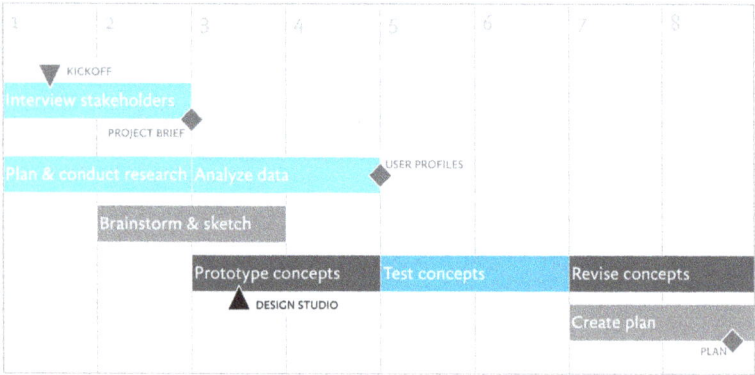

**FIG 4.5:** The basic plan adjusted to prepare for prototyping activities with research, yielding a prioritized set of feature ideas.

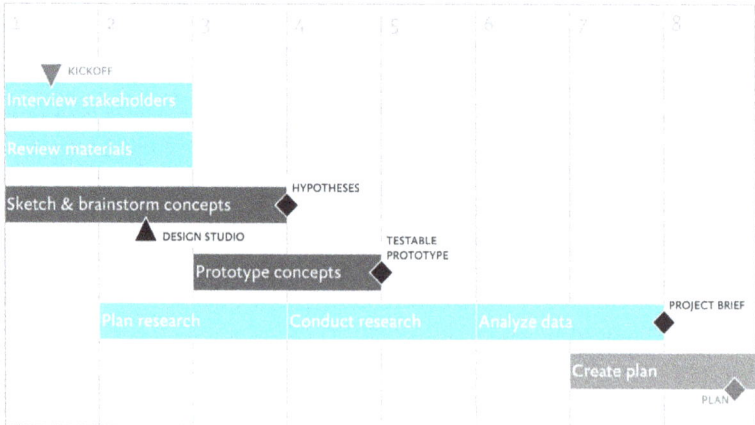

**FIG 4.6:** Lead with a brainstorming session following stakeholder interviews to highlight the big questions that need to be answered through research.

Don't get sucked into designing whatever you brainstorm, though. Generating pictures early can be valuable, but only if you don't compromise the rest of the design process. Use the design studio to highlight the range of remaining decisions and the unresolved requirements for the product. Even if it's just a few things, use these as an excuse to conduct some research so you can start refining a direction.

## WHEN TO STOP, WHEN TO KEEP GOING

You may get to a point when discovery isn't yielding any additional value. In the language of this book, the discovery mindset is no longer moving the project forward. Perhaps it's time to switch to a mindset where you're making more decisions about the user experience itself. Be on the lookout for these warning signs:

- **No new insights.** Whether you're sharing observations about the target audience, putting business requirements into context, proposing a new big idea, or defining phases in a roadmap, you're making assertions as a direct result of discovery. At some point, you may not be able to say anything new about the problem itself or the direction to go in. Sure, you've got the details to figure out—button labels and precise content categorization and the nuances of a transactional flows. But these tasks don't say anything new about the assignment. This is the surest sign discovery has come to an end, for now.
- **Increasingly granular insights.** Sometimes you gather so much data and extract so many good principles that you start digging deeper—for instance, you segment the data and notice that one group of users has a slight preference for product benefits at the top of the page, and the other group prefers compatibility information. You have to ask yourself, *How much does this really help the design team?*
- **Impatience to flesh out details.** There comes a point when you're so excited about the idea, you just have to see how it plays out. This seems pretty subjective, but it's not a voice to ignore. You shouldn't deny yourself the satisfaction of elaborating the details, so long as you can make decisions confidently. As soon as you find yourself leaning on assumptions, ask yourself, *Do I have enough information?*

- **Feeling anxious about elaborating the design concept.** Perhaps you're not so much excited about working out the details—what users see when they click a certain button, or how an animation works—as you're nervous about what you'll find when you do. This is also not a voice to ignore. Start working through the details, but be vigilant: you might be going down a rabbit hole to quell an irrational fear. Give yourself permission to acknowledge you might have more learning to do.

On the other hand, if you're feeling hesitant, you may not have enough confidence to move forward. You may need to switch gears and start exploring concepts, or step back and think about the information you're missing. There are other signs that you're not done with discovery activities just yet:

- **Debating without resolving.** Through discovery, design teams build a framework for making decisions. Imagine you and a teammate are arguing about content priorities—which content should go higher on the page. Or perhaps you're arguing about which features are essential to your minimally viable product. If you can't make a decision, ask yourselves what information you need to facilitate that.
- **Lack of inspiration for design.** Discovery gives designers the spark of inspiration. That spark heats the design process until the end—it is the reason you're doing the project at all. If you haven't identified that spark, you may be focusing on the wrong things—perhaps too much user research and not enough brainstorming, or too much brainstorming and not enough planning. All activities are essential for igniting that spark. Step back and look at what you may have been neglecting, so you can focus your discovery efforts.
- **Discomfort with the unknown.** Discovery empowers designers to venture confidently into the unknown world of detailed design. Discovery lights the way. Wariness of the unknown could signal many things—anxiety about working out the details, for example—but it could also mean you need more time to explore those questions.

# DEALING WITH DISCOVERY OBSTACLES

Discovery gives your team an opportunity to consider different directions, permission to ask difficult questions, and freedom to try new approaches to the same old problems (or common approaches to new problems). But with these things—opportunity and freedom and novelty—comes the risk that your team may lose themselves in this trek into the unknown.

Good planning means accounting for risk. Keep an eye out for the following situations.

## When you ignore the goal

Discovery projects generally represent a departure from the day-to-day activities of a designer's usual responsibilities. User research, brainstorming, and conceptual thinking are, for many team members, oases in otherwise dreary workdays. It's therefore tempting to do these fun and exciting design activities without regard to their purpose.

Stay pragmatic. Remind your team that putting ideas into practice is the ultimate goal. Remember to provoke difficult conversations. Sometimes asking questions about why something is important can help you show how it relates to the day-to-day.

## When everyone thinks they're right

You may find that stakeholders have preexisting ideas about the product. Maybe they have strong opinions about what users think or need. Or they insist they know what content is important.

In these situations, the project seems less about exploring new territory and more about validating ideas. These project team members don't want you to find something new—they want you to prove them right. I've come into some projects where the project team has divided itself into two camps, each with an idea of the right way to design the product.

In this situation, make sure you're answering explicit questions. Design your projects to directly address the team's conflicting perspectives. Translate conflicts into testable hypotheses. Conduct brainstorming sessions that force team members to elaborate on their positions and understand other viewpoints.

You can also use discovery to highlight the assumptions behind the preconceptions and gather data to debunk them. Help the team see the problem in a new way. Emphasize strong user insights to encourage people to consider the product from a new perspective.

## When everyone has different expectations

Aligning project expectations requires ensuring everyone on the project understands the outcome. For one project, my discovery work involved three user research studies, initial concepting, and validation testing. One stakeholder was concerned that we didn't arrive at final, implementable designs in that time, despite frequent level-setting conversations. "Where are the screen designs?" they asked. "Why can't we just build this?"

Before you present your work, think about the arguments against it. *What went wrong in the past? What did they ask about or seem concerned about?* Draw on your experience and anticipate their concerns.

When you're presenting, frame the conversation so everyone on the team understands the decisions you're working on and how they relate to the larger product experience. You may be talking about specific content, but it's an example of the overall tone of the site. You may be talking about brand attributes, and those will have very direct implications for subsequent visual decisions.

At every meeting, remind participants of where you are in the project process, what you've done so far, and where you're going. Convey status, be honest about the problems you're facing, and remind everyone where you hope to end up. You might say, "I'm keeping an eye out for X," or, "Last time I did this, we had trouble with Y." Tie the process together so everyone's on the same page.

## When you face tight deadlines

When you're short on time, inexperienced stakeholders may want to drop discovery altogether: "Can't you just start designing?" they ask. But hopefully they understand you need to understand the problem before you can solve it. Good ideas take time.

There are all kinds of ways to compromise on time. The best way to keep your research, analysis, and processing time short is to prepare specific questions at the outset. What do you need to know to set a direction? Pick three questions, then gather enough data to answer those questions.

Give yourself at least half a day to go through the insights and find a few themes or patterns. Once you've established your project goals, start drawing and getting feedback right away. It may seem like you can save time by working alone; while it's true that a group is more likely to be tempted by detours, ideas are better when vetted by multiple minds. Try brainstorming and workshopping ideas in smaller groups—three to eight people is a good size, depending on the activity. Establish a ground rule that you need to keep each other honest and on-task.

When you start focusing on a direction, present two options to choose from instead of iterating and refining together. The time savings will likely affect the quality of your work, but may convince your team to make room for more discovery later.

## When people's agendas don't align

If members of the team aren't aligned—or are verging on adversarial—their motivations and involvement with the project can be so different that they seem to be from two different planets. You may not see it at the outset, but if you do, you can structure discovery to deal with the situation.

Start by interviewing stakeholders separately. Learn about each position, motivation, and agenda without adding to the conflict. Find the crux of the problem. Is it personal, territorial, or just a misunderstanding? For example, if your team disagrees about the structure of the site navigation, perhaps they're really dealing with conflicting business priorities, or different ideas about the target audience.

Find some common ground in the interview results, like concerns about the users or a frustrating business process. Meet with the stakeholders together to review these divergent and convergent areas. Choose a strong facilitator to lead this session—they can encourage discussion around conflicting viewpoints and help stakeholders hash out their differences in pictures. Wrap up by summarizing where they come together and where they differ.

With these activities behind you, you may see the disagreement in a new light. Try to reframe the disagreement as a hypothesis for testing, or in terms of the target audience. Help your team put it in perspective.

Each of these tweaks has a negligible effect on the overall plan, but may require a small amount of time in certain activities. Avoid becoming consumed with internal politics. The project will benefit from discussion, but shouldn't be defined by conflict.

## When there's no strong leader

Discovery is collaborative, but you still need a strong leader: someone to set direction and develop a clear vision of what the product should be. This leadership manifests itself at multiple levels, from the day-to-day prioritization of tasks all the way up to the positioning of this product relative to the rest of the portfolio.

When leadership is lacking at any of these levels, projects falter, especially in discovery. Without a clear product owner, you'll be trapped in endless cycles of non-decisions. Lack of leadership exposes the truth about discovery: no process can answer every question.

A lack of leadership also means managing expectations and conflicting agendas. There may be an overriding distraction of keeping everyone happy, a focus on building consensus instead of building what's right. Getting the organization aligned is one thing, but keeping stakeholders happy is another.

As team lead, your responsibility only goes so far. You may not have the budget or power to decide what's important for the larger organization. Some strategic decisions may rest with another group, like marketing or brand strategy. Regardless of your level, you'll likely encounter a lack of leadership at some point along the way.

Perhaps the hardest thing to do in this situation is acknowledge that it exists. In some cases, it's obvious: there's a vacuum of leadership because there's no one in the position to make a decision. ("We actually don't have a chief marketing officer!") Other times, it's someone not doing their job, or not having a vision.

Articulating, at least for yourself, the decisions you need someone to make is the first step in dealing with this situation. Lay out the choices and the consequences for each of them. For instance, removing a feature may be a reasonable decision, but it could make your product less appealing to the target audience, or affect the morale of your team.

Once you've explored the options, you may need to adopt a more assertive position and set the direction yourself. The common saying to describe this attitude is, "Seek forgiveness, not permission." Go forward with decisions that are justified by the information you have, knowing you may need to explain them later.

## When people don't understand design

Let's assume the worst: you are forced to work with a team that thinks of design as making wireframes pretty—not an ideal project for any designer. Designers like you and me want to make meaningful contributions; applying a color palette and adjusting typography are meaningful decisions, but only insofar as they contribute to something bigger.

When you question this team about their methodology, they will respond with hostility. People unfamiliar with design will respond to questions about discovery with "We didn't have time for that" or "We already knew all that" or "This project isn't that complex." You will be tempted to prove them wrong. But you're not here to be right; you're here to help, and clarifying their vision is the best thing you can do.

To avoid spending the next six months banging your head against a wall, learn as much as you can about the proposed direction and the decisions that went into it. Run usability studies to assess possible gaps. Help the team understand what's missing from the direction, why these elements are important, and what steps you can take to fill the gaps.

Get people to explain their decision-making process. If something is confusing, ask them to frame it as a story with a real user in a real situation. When they make inferences about user intent or behaviors, ask them where they got those insights and press them on different scenarios. Compliment their ideas with phrases like "That's really good" or "That's an interesting new approach," then follow them up with "I'm curious how you got there."

Consider different ways to extend, enhance, and elaborate the proposed vision that might address some of the gaps. Ask for help as you work to extend the direction, and highlight things that remain unclear. When you demonstrate genuine interest in the project, you position yourself as part of the team. You create opportunities to work together, and help your team understand how design works.

With each of these techniques, your tone is crucial. You can ask a question with sincere curiosity or an underlying desire to undermine your teammate. You're not trying to corner them, but instead trying to find opportunities to get involved.

## SETTING THE RIGHT TONE

Picture two projects with the same objectives and similar teams. In the first project, the team is on the defensive from the start. They don't believe the project stakeholders will play nicely. They're already lamenting the constraints, which seem especially unreasonable. And they convince themselves that the project itself is boring. In the first meeting with the stakeholders, the collective team doesn't gel. They go through the motions, checking items off the meeting agenda. The designers roll their eyes when one stakeholder pulls out printouts of wireframes he made in PowerPoint. The stakeholders get exasperated when the designers show their ignorance about the domain. No one is excited about the project. The design team walks away feeling as if their worst fears were confirmed. The stakeholders walk away ready to believe every bad thing they've ever heard about designers.

Now consider a different scenario: The designers come to their kickoff with a list of questions. One of them facilitates the discussion while the other captures information on the whiteboard, only interrupting to make sure they've heard things correctly. They don't get defensive when one stakeholder pulls out wireframes, and instead ask them questions about their thought process. They ask questions about the domain, treating the stakeholders as subject matter experts. Their questions come from ignorance, but also convey that they've done their homework. In this case, the teams depart with heightened enthusiasm and confidence. Worried less about their egos or facing the inevitable challenges, these teams take the opportunity to understand and solve a problem.

Great discovery depends on well-structured projects and embodying the attitude. This means acknowledging that you don't know the answer to every question. It means giving people the benefit of the doubt, because perhaps they can see the problem in a way you can't. It means relishing the opportunity to look at the design challenge, your team, and yourself in a new way.

Embracing this mindset gives you a chance to set the tone for the project at the outset. Humility and curiosity go a long way toward helping the team work together, fostering healthy conflict, and keeping things moving.

These are the postures I adopt, especially when working with a new team:

- **Ask for help.** When you're working on something, don't hold back on asking for help until the very end. Discovery is about rough ideas and refining your understanding. Use the people around you to vet, validate, and elaborate.

- **Ask for tasks.** When you're nearing the end of the task, make sure your lead knows you need something new. Better yet, volunteer activities you think are appropriate given what you've learned so far.
- **Ask a lot of questions.** If you've run out of things to ask for, ask a question. When someone is talking, listen and think about what to ask next. Write down keywords and use those to ask for clarification or more detail. Keep an ear out for assumptions, and ask people to tell you more about things they've said. The hardest part of this is tricking yourself into asking a question instead of responding. Ultimately, you're trying to cultivate (and model) behavior that encourages people to question underlying assumptions and avoid taking things for granted.
- **Avoid defensiveness.** When people ask you questions or offer feedback, assume they're coming from a position of wanting to help the project. Engage in a conversation, not an argument.

## PLANNING THE UNPLANNABLE

More than any other aspect of the design process, discovery follows how people learn: you can't gather information and expect to be able to use it as-is. You need time to apply what you learn in different ways, reflect on it, and challenge your assumptions. Project plans help you plan and track that time.

But discovery never goes as planned.

Modern product-development methodologies put an emphasis on speed—time-to-market, failing fast, rapid prototyping, sprints. But innovation doesn't happen overnight. Discovery, at least as presented in this book, can be as fast or slow as you like. Methodologies and frameworks come and go, but we'll always need to fill gaps in our knowledge—we just have to find the time.

Despite the evolution of modern web development, methodologies, project plans, deliverables, and milestones continue to play a role in structuring our work. The next chapter deals with making discovery real: planning activities and preparing deliverables.

# 5 DOCUMENT THE OUTCOME

> *The mess actually represents the deep and meaningful sensemaking that drives innovation.*
> —JON KOLKO, "Abductive Thinking and Sensemaking"

OVER THE YEARS, I've used dozens of different methods to document what I've learned in discovery, including personas, mental models, requirements, journey maps, storyboards, and flowcharts. The format of these deliverables varies depending on the project, and, frankly, current trends for documentation. In the early days of web design, teams relished elaborate design documentation. Later, we rejected formal deliverables. And now? We're seeking a balance—formality when needed, informality otherwise, and neither at the expense of genuine communication.

Perhaps the best argument against discovery documentation is that teams get value from actively participating in the process, not reading about it. So why document? The truth is, not everyone can participate the entire time, and different people interpret things differently. Documentation is an insurance policy. If your project budget or schedule limit who can be involved, documentation can help you bridge the gaps. And if your team has conflicting opinions about priorities, documentation can help your team rally around core elements.

A good discovery document serves as a foundation for all of the conversations you'll have in the future. Comprised of insights about users and business objectives, design principles, and validated design concepts, this foundation serves as a reference and a shared vision. With a little planning, you can use it to justify your design decisions, so long as you draw meaningful connections between the work and what you've learned up to this point.

## THE ESSENTIAL DISCOVERY DOCUMENT

Whether you call them discovery briefs, creative briefs, or design briefs, these documents have long been part of design vocabulary. But with everything it covers, this deliverable isn't always as concise as its name implies.

A typical brief captures:

- objectives and outcomes
- the problem statement
- the design direction
- supporting details
- next steps

For the rest of the chapter, I'll refer to the discovery brief as *the document*. But you'll choose how formal, detailed, and complex it is—and you can decide what to call it. You may share a handful of notes, a hand-drawn poster, or a sixty-page PDF. You may create a clickable prototype and support it with annotated screenshots. Make it fit the project and the amount of work you did.

## Objectives and outcomes

Resist the temptation to document discovery as a chronology. Like most good business writing, your discovery document should begin at the end. Or, to use another cliché, give me the "so what." That might feel like you're not saving all the "good stuff" for the end, but this document (or really any design document) shouldn't be a slow reveal. As a guide for the remainder of your project, it needs to set the tone. Therefore, start with not only what challenge you're trying to fix, but also what you'd like the outcome to be, and what will guide your path there.

Remember the assertions from Chapter 2? They included project objectives—various ways of capturing how the project will address some difficulty in the world—that may be expressed as tangible outcomes, or as a mission or vision. You can also draw upon the assertions described in Chapter 3, stating the direction you'll take, and perhaps summarizing the guiding principles. Use the first few pages of the document to describe those assertions and the overall purpose of the product—not just what you did in discovery (FIG 5.1).

You can also set up the document by describing how the discovery process supported your objective. How did you approach discovery? What have you done so far? Summarize the activities, milestones, and outcomes, either in a Gantt chart or scannable list (FIG 5.2).

## The problem statement elaborated

Though you introduced the problem statement in the overview, you learned a lot about this problem through discovery. A thorough explanation will be helpful later in your design process. So the next section of your discovery document should elaborate on the problem statement, primarily by relying on those contextual statements. This is your opportunity to highlight what you learned about the business and users and technical environment and content ecosystem—at least, to the extent they have an impact on your project.

Perhaps the most straightforward approach is to dedicate a page in your document to each gathering activity. The page for

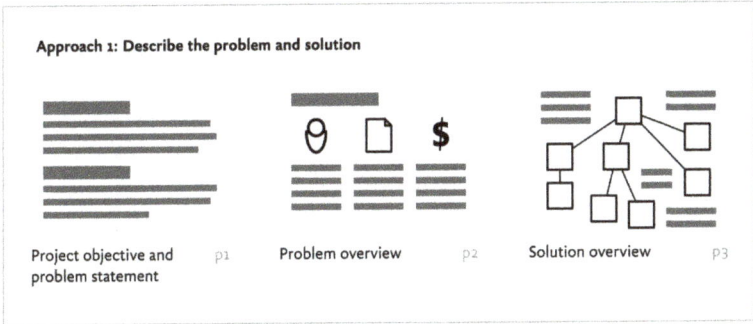

**Approach 1: Describe the problem and solution**

Project objective and problem statement    p1      Problem overview    p2      Solution overview    p3

**FIG 5.1:** In one approach to summarizing discovery, you lead with the outcomes themselves.

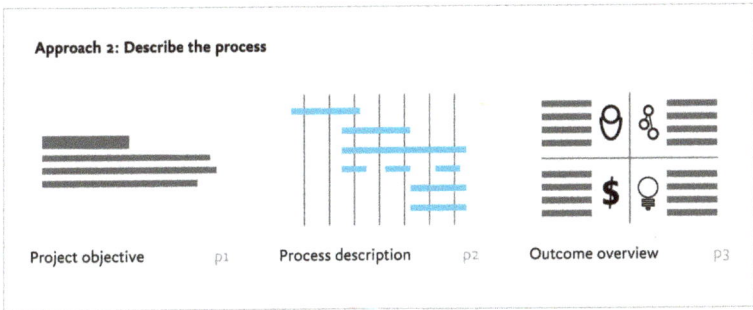

**Approach 2: Describe the process**

Project objective    p1      Process description    p2      Outcome overview    p3

**FIG 5.2:** Another way to summarize discovery: describe your approach for discovering the solution and explain the conclusions drawn from that process.

user research should, for example, summarize your insights from the interviews you did. The page for the competitive landscape review should highlight things you learned looking at the competition (**FIG 5.3**).

Another approach is to take each theme—the top insights and conclusions you drew—and elaborate on each of those through what you learned (**FIG 5.4**). This works particularly well if you've used the "Find the patterns" technique I described in Chapter 2, which helps you pull out threads from the research. This document is your opportunity to weave those threads together.

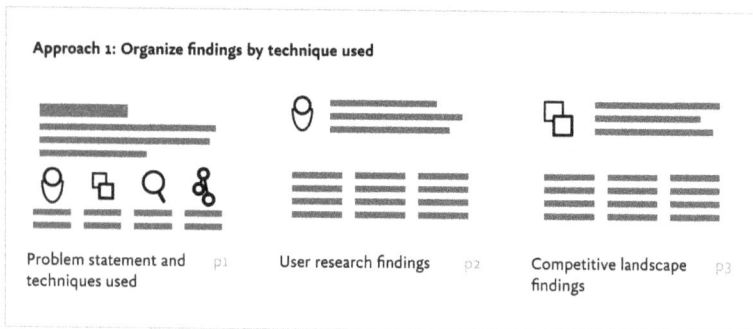

**Approach 1: Organize findings by technique used**

Problem statement and techniques used p1

User research findings p2

Competitive landscape findings p3

FIG 5.3: Capture the problem by describing each discovery activity and the findings that emerged.

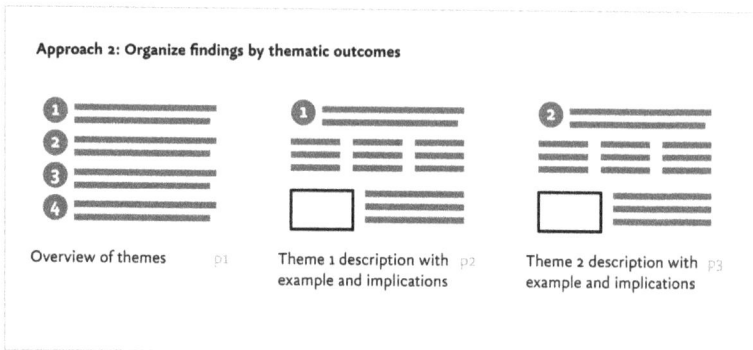

**Approach 2: Organize findings by thematic outcomes**

Overview of themes p1

Theme 1 description with example and implications p2

Theme 2 description with example and implications p3

FIG 5.4: Alternatively, organize the problem description by highlighting themes and what they mean for the product.

## The design direction

In the next section, explain how you'll solve the problem. As with the previous two sections, you'll draw upon assertions, primarily the principles and models from Chapter 3.

It may be tempting to dig into specific screens right away, but remember you're documenting for your future self: what context do you need to understand the direction? One way is to start with the *big idea* or *central concept* assertions: represent

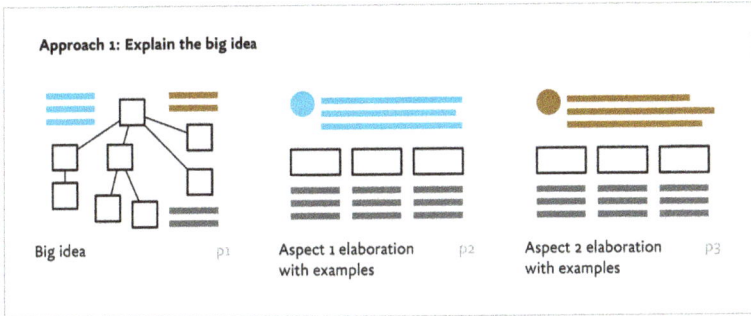

**Approach 1: Explain the big idea**

Big idea    p1     Aspect 1 elaboration with examples    p2     Aspect 2 elaboration with examples    p3

**FIG 5.5:** Describing the design direction depends on the techniques you've used to define it. If you have a big central concept—a primary structure or metaphor—you can show that and elaborate on individual pieces.

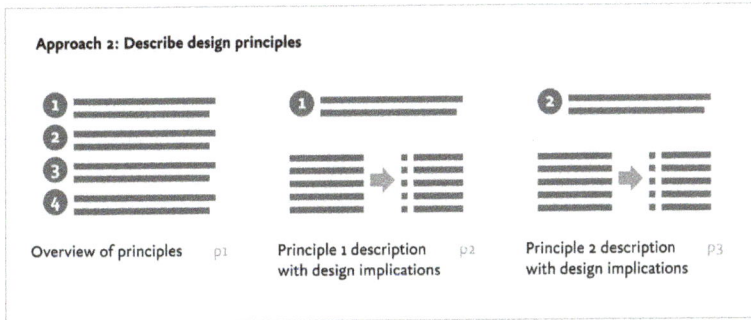

**Approach 2: Describe design principles**

Overview of principles    p1     Principle 1 description with design implications    p2     Principle 2 description with design implications    p3

**FIG 5.6:** If your design direction comprises a set of principles that serve as the guiding force in defining the product, you can elaborate on each principle.

the idea that's holding the whole thing together. This framework then sets you up to dig into specific aspects of the big idea (**FIG 5.5**).

Another way is to focus on the principles (that *other* assertion from Chapter 3)—the rules that will guide you up the mountain (**FIG 5.6**). Recall that principles will help you shape the design, and therefore set a design direction, helping stakeholders picture what's important.

In either case, be sure to support the big idea or design principles with visual examples. Whether you're drawing from your initial design ideas, sketches, competitive examples, or the existing product, pictures will help you remember what you meant by a principle or implication. They can help make big ideas concrete.

To make sure everyone is working from the same starting point, you can include some design basics here, too. At my design firm, we occasionally touch on the fundamentals of color theory or information architecture to educate stakeholders and facilitate the decision-making process.

## Supporting details

Whenever I'm assembling a discovery document, my primary goal is to present a cohesive story. By this point in the process, I've learned so many things and made so many different decisions—and I want the project team to understand how everything fits together.

The details *do* matter, but because you've spent weeks or months immersing yourself in them, it's hard to know which to include. The relevant details are the ones that best support your conclusions.

There are two ways to organize the document to elaborate on the details. In the first method, you explain the results coming out of each of your major design activities. If, for example, you did baseline usability testing, a content audit, and visual design analysis, the document could have a section on each of those (FIG 5.7).

Many of my discovery documents start out this way, as a scaffold for capturing everything we're doing. The document evolves, arriving closer to the second approach, in which I dig into each design theme or aspect of the product concept. With some products, for example, it's easy to think of them as a set of major features—navigation, search, content display. You can elaborate on your findings for each feature in subsequent sections of the document (FIG 5.8). For less literal product concepts, you can break down the details in terms of scenarios or user types.

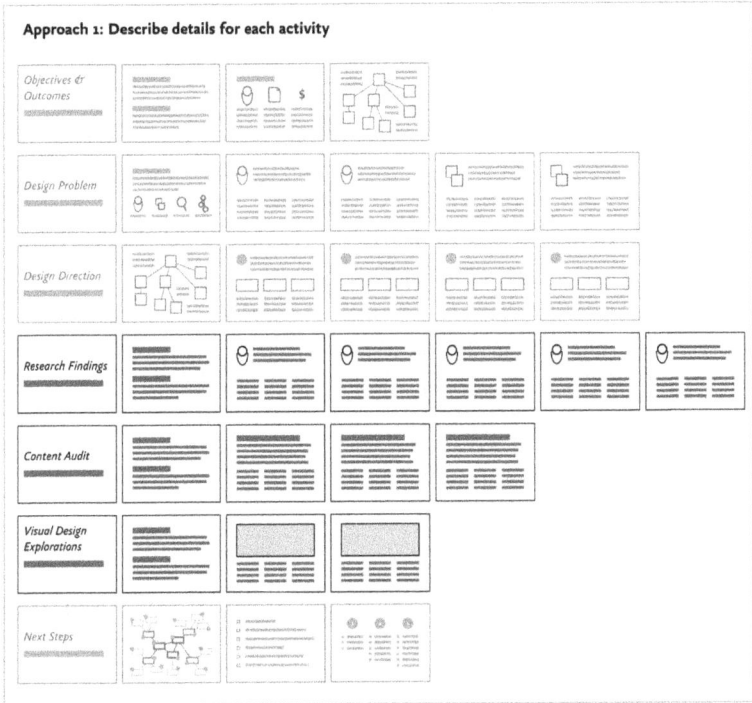

**Approach 1: Describe details for each activity**

- Objectives & Outcomes
- Design Problem
- Design Direction
- Research Findings
- Content Audit
- Visual Design Explorations
- Next Steps

**FIG 5.7:** I use diagrams like this to storyboard documents: rectangles on the far left show section headers, while rectangles to the right show different types of pages I'll use to tell the story. In this approach, after introducing the problem, each of the main activities from discovery gets a section explaining the outcomes of research, audits, and brainstorming.

Conclusions from discovery can seem abstract, so the details make them more human. For a theme like "When picking a career, high school students prioritize income to justify their college costs," quotes from students make this more than just a design principle. It turns it into a way to be empathetic with the target audience.

These kinds of details will flesh out your narrative:

- **Quotes or behavior observations** from users and stakeholders make for great illustrations. Line quotes up with findings or themes to drive home the point. Experience

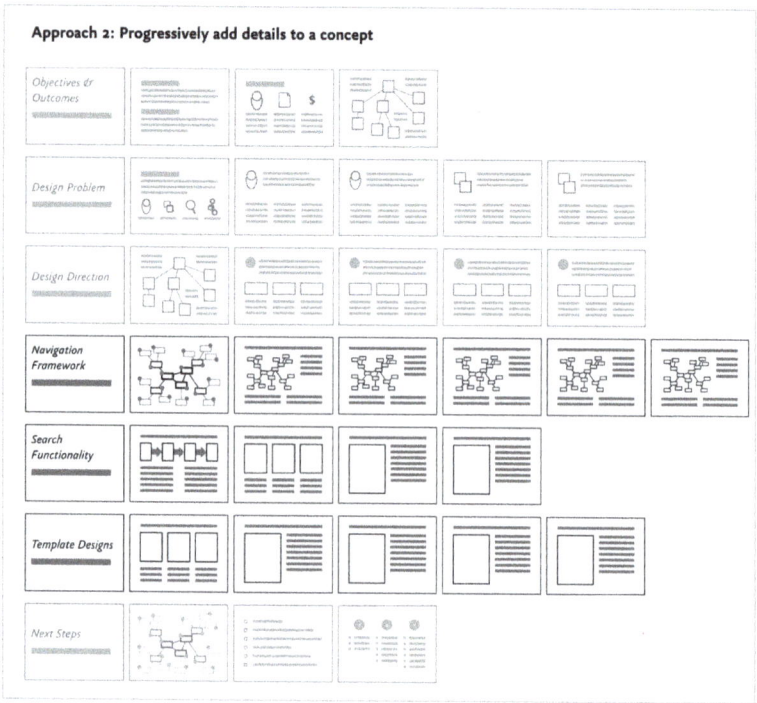

**Approach 2: Progressively add details to a concept**

**FIG 5.8:** Another approach is to reveal more and more of the big idea—looking at the product in terms of different features or perspectives or usage scenarios.

has shown me that stakeholders gravitate to these verbatim statements—they'll come up again in subsequent conversations—so use them judiciously. Too many quotes and they're difficult to remember.

- **Quantitative data**—like graphs and statistics with clear takeaways—can be very powerful. Increasingly, I'm using quantitative tools like Optimal Workshop's Treejack and Chalkmark (http://bkaprt.com/pdd/05-01/) to show the relative success of certain aspects of the user experience. These tools render results in bar graphs and pie charts, perfect for your documentation.

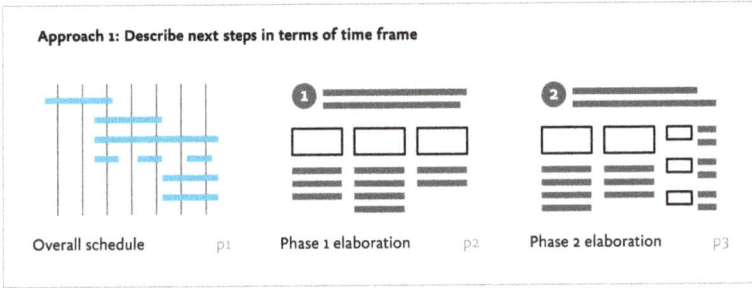

**Approach 1: Describe next steps in terms of time frame**

Overall schedule    p1      Phase 1 elaboration    p2      Phase 2 elaboration    p3

**FIG 5.9:** Your team may have a favorite way of planning projects, so use the method that will resonate best. I've created plans that detail how the product can be built over a series of phases.

## Next steps

Wrap up with a plan for fleshing out the design. You may want to prioritize features or aspects of the product based on your research. Complex features need more time, and you may have a good understanding of what's complex based on your brainstorming and focusing. Your plan may also depend on other parts of the organization—engineering, marketing, or customer support, for example. Make a note of these dependencies, even if your team is still planning them out.

If you have a sense of the time line, map it out in weeks, months, or quarters. If it's difficult to commit to a specific time frame, use phases or sprints. If you're not sure how you intend to structure the remainder of the project, you can, at the very least, highlight major milestones in the product's development (**FIG 5.9**).

While a time line is an obvious approach here, you may not be ready to commit to deadlines. In these cases, your discovery efforts may conclude with a list of potential projects (**FIG 5.10**). For example, for my search engine project, I developed a menu of feature ideas, including advanced filters, personalized results, and suggested matches. If you're outlining potential projects, explain the value, benefit, level of effort, and likely costs for each of them. This will help your team understand their options.

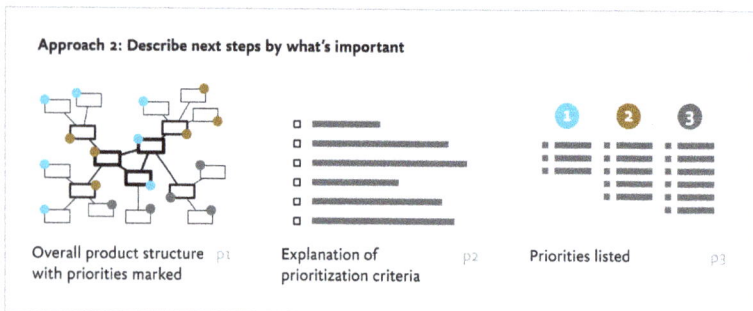

**Approach 2: Describe next steps by what's important**

| Overall product structure $_{p1}$ with priorities marked | Explanation of $_{p2}$ prioritization criteria | Priorities listed $_{p3}$ |

**FIG 5.10:** I've also been less committed to a time line, revealing instead which features should be built first.

## DOCUMENTATION BEST PRACTICES

The previous section gives you a starting point for the structure of your discovery document. Creating great documentation requires awareness of the context and the application of tried-and-true practices, just like design itself. Keeping these guidelines in mind can help make your documentation more effective.

### Think about your audience

While it's worth sharing more broadly, discovery documentation is for the product team. Its central purpose is to inform and inspire future design work. Don't try to turn it into a sales tool. If you need to justify the cost or effort involved in the project, create a separate document for people who hold the purse strings.

At some point before the table of contents starts to take shape, try to anticipate what you'll need in the future. What kinds of questions will you have when you're designing the details? This requires extraordinary forethought, and it's hard to get it right—but it's important to try. Ask yourself questions like:

- What do our users care about?
- Of the things they care about, what's most important?

- What is the one part of the product we have to get right?
- Is some information more important to the business than it is to the user?
- How can our voice and tone support the user's needs?
- What will turn users away?
- What are great examples of representing the organization's brand?
- Which technical constraints have the greatest impact on the user experience?

Anticipate the kinds of decisions you'll have to make later, which are the same kinds of design decisions we have to make on any project. Imagine the kinds of conversations you'll need to have. Write the documentation for that future version of yourself, who needs to make those decisions and have those conversations.

## Make it concrete and actionable

Discovery spends a lot of time in the abstract, the imaginary, the blue sky. You've already taken your team into new territory. Don't get too philosophical, or distracted by every novel connection or insight, or you'll risk losing your team's attention. Use visuals to make your ideas more tangible (FIG 5.11-13).

Ultimately, to be understandable, your document must surface the central claims. Don't bury them in a report. Give that future version of yourself a one-pager of design principles, so they're all in one easy-to-find place. Make sure you have the product vision articulated and supported succinctly, so future you can turn to it throughout the project.

## Start informal

Whether or not you intend to prepare something formal, start informal. Designers often face harsh criticism when they focus on the deliverable rather than the outcome. Starting informal means you can evaluate the utility and approachability of the documentation. Create a detailed outline. Review it with your team. Spend time talking about the messages, assertions, and

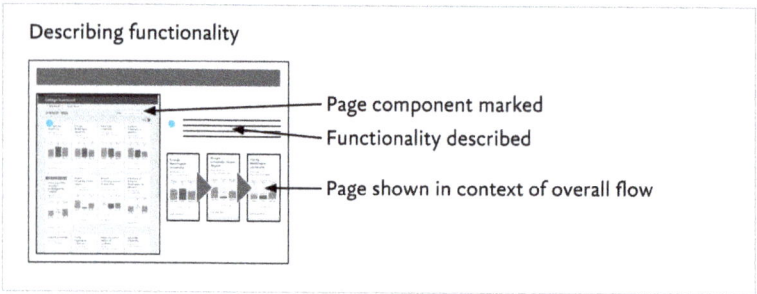

**Describing functionality**

- Page component marked
- Functionality described
- Page shown in context of overall flow

**FIG 5.11:** Describe functionality by showing how different screens relate to each other across the product.

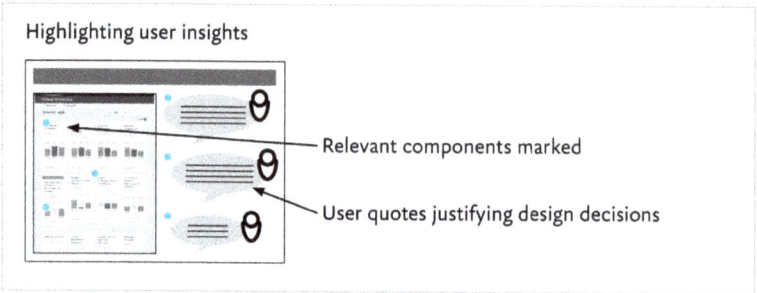

**Highlighting user insights**

- Relevant components marked
- User quotes justifying design decisions

**FIG 5.12:** Show how user insights translate to design decisions by aligning quotes to specific aspects of the design.

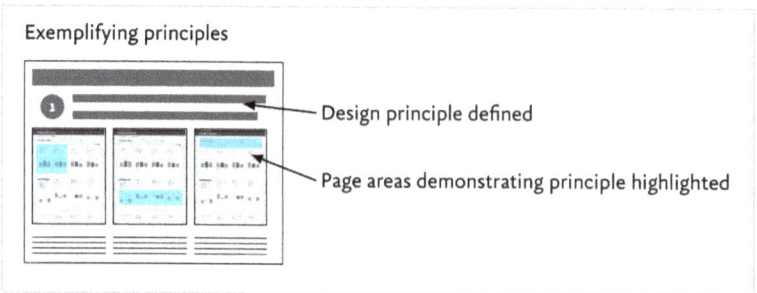

**Exemplifying principles**

- Design principle defined
- Page areas demonstrating principle highlighted

**FIG 5.13:** After stating a design principle, show examples of how it influences the screen or page design.

themes to get them at the right level, and make sure they make sense. Don't try to perfect it on your own.

## Frame the story

It's tempting to reveal the big idea at the end. You build up the narrative, constructing an argument that leads to an inescapable conclusion—or so you hope. Unfortunately, this doesn't always work in practice.

Create a structure that starts with the conclusion and central claims. Reveal your intent right up front, with a summary of the design direction, and then use the rest of the document to support it. The introduction is where you set the tone and capture the big picture. It's your opportunity to set the stage and frame what's to come.

Use the structure of the document to emphasize the important assertions. In creating a structure, you're driving the narrative and setting up future conversations about the project. Your document should conclude with next steps, demonstrating that you've planned how to execute the project and you acknowledge this is just the beginning of the conversation.

## Focus on themes

It's also tempting to capture every finding or constraint, but even the shortest projects yield too much information to put into a digestible document. Focus on the most important themes from your research and exploration. You might call these high-level findings, key themes, or design principles. Summarize the details, but don't try to replace them.

In user research, you may want to capture every pattern or exception. Yet sufficient analysis often reveals half a dozen or so themes. Good discovery documentation focuses on the essential parts of the story: the observations, principles, and concepts that will help you move forward. Skip things that don't directly relate to the project, and sum up insights that point in the same direction.

### Tie it to a specific problem

One litmus test I use for my own documentation is to find-and-replace the project name. If I can safely swap the client's name for another without losing meaning or truthfulness, I'm not saying anything particularly insightful. Some industry adages are necessary to lay the groundwork (for example, good search experiences depend on good content strategy), but the bulk of your document should focus on your team's particular needs, expressed in language they can relate to.

### Use consistent language

One of my favorite consequences of discovery is inventing a language. It's not Elvish or Klingon or whatever they spoke in *Avatar,* but you're still working with a shared vocabulary. You need to label different aspects of the product, like *features, user segments,* and *content types.* You need to invent structures that tie the user experience together, and those structures have components that also need labels.

In doing work for a search engine, I needed to distinguish what the search engine found ("matches") and the way it was displayed ("results"). It was important to use the terms the same way every time to help the client understand the distinctions. It put a damper on my writing, but it helped me make sure everyone was talking about the same thing.

### Don't think of it as "final"

Remember that this is a foundation for the work ahead—not a contract or design specification. This document needs to align teams, not stand the test of time. You may need to revisit your assumptions or decisions in the future. Good documentation evolves with the process, showing a snapshot of where you currently stand and helping you build consensus as you move forward.

## THE BENEFIT OF MAKING THINGS REAL

I started this book by describing the discovery mindset, an attitude that acknowledges a lack of knowledge or understanding. Embracing this mindset means diving into the unknown and channeling all the curiosity, skepticism, and humility we can muster. The process is the thing.

By engaging in the activities described in this book, you and your team internalize a new understanding of the design problem and the context your product will live in. You can confidently choose a direction, knowing it will address the need.

How does documentation fit into all of this? Documenting our effort makes it real. It allows us to confirm that our understanding of the problem is mutual, that the principles driving design are well understood—and that we won't forget them. It gives us a vision we can believe in—something we can work toward building together.

# CONCLUSION

THERE IS NO DESIGN without discovery.

As soon as you decide to solve a problem through design, you're deciding to learn something—even if you don't plan to conduct any research or test your ideas. The moment you start drawing, you're seeing the problem anew, and you start asking questions: *Which element is more important? Do users need all this context? What does the business lose if we change the inputs on this form?*

By putting ideas down on paper, you've already triggered your curiosity, a need to ask questions and fill gaps in your knowledge. The very act of taking steps to solve a problem prompts you to ask yourself, *Will this work?* The activities and techniques in discovery didn't emerge by accident: they have withstood the test of time because they give designers tools for adopting new perspectives, asking the right questions, and finding useful answers.

For all our talk of techniques, activities, and deliverables, discovery is still a mindset: it's a perspective you adopt to define a problem, then fix it.

Design is a creative endeavor because it relies on our ability to connect ideas to develop novel approaches to solving problems. Many great thinkers have sought the secret to creativity. What makes a person or process creative? What environments are conducive to creativity? What happens in adulthood that takes away our childhood ability to imagine new things?

Research into creativity offers no satisfying answers, but instead gives us hints about what works. Build up domain knowledge. Give yourself time to make connections. Collaborate with other people to encourage mixing ideas. The history of innovation is filled with stories that are more about the long cultivation of knowledge and percolation of ideas than some eureka moment.

It's this history that we hope to participate in, the history in which we find major advancements like the invention of flight or the polio vaccine or the iPod. We want to design products that are elegant, that fit effortlessly into someone's life, and perhaps change it for the better. Discovery's tools and techniques, coupled with the mindset, give us a means for dipping a toe into the history of invention. Discovery bridges the gap between creativity and business.

For all our efforts to recreate the conditions of great invention, design happens in the confines of a project, something with milestones and a responsibility assignment matrix and a budget. If you've fallen in love with stories of invention and analyses of creativity as I have, this might make you a little sad. The modern pace of business demands mechanizing an otherwise natural process to align with corporate appetites. The world of business thrives on deadlines and predictability, and measures itself on efficiency and profitability. With the mounting pressure of competition and market scrutiny, businesses are ever more preoccupied with getting something "out there."

It's difficult to reconcile this urgency with what we know about creativity. And that's what discovery is: a compromise, the lovechild of business process and cognitive process. It's what happens when we subject creativity to the opposing force of business—a means to structure it, and make it more predictable, while preserving the value.

But what does this mean from the designer's perspective—from your perspective? Are you solely responsible for bringing a creative mindset to the business? It's hard enough to do our jobs, much less influence corporate culture. You may be tempted to defend design's methods or outputs. You may insist on doing user research or multidisciplinary brainstorming. You may become frustrated with compromising your time to synthesize and reflect, or with shoehorning insights about users into a specific format. You may choose to step up and fight the good fight, if not for the sanctity of your process, then for the benefit of your users.

But, before you steel yourself against attacks on design process, ask yourself, *Does this help my team understand the discovery mindset?* The best way to make a change is to take it slow and exemplify it. Carving out a role for discovery in your company means showing how the mindset can be woven into things your organization already does. It means helping people understand that design isn't a rote process, but instead is a way of gathering information, putting that information into context, exploring ideas, and focusing on a direction.

The best way to help your colleagues understand discovery is to do it—to show them that sometimes the best solution lies not in a process or a deliverable, but in an attitude.

# APPENDIX: THE COMPLETE DISCOVERY ACTIVITIES MATRIX

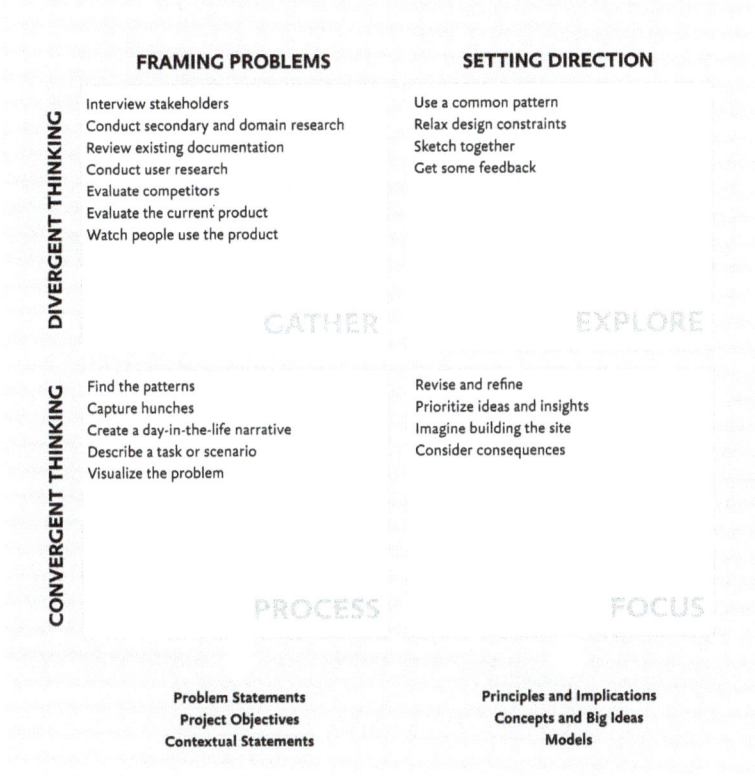

**FRAMING PROBLEMS**                    **SETTING DIRECTION**

**DIVERGENT THINKING**

Interview stakeholders                  Use a common pattern
Conduct secondary and domain research   Relax design constraints
Review existing documentation           Sketch together
Conduct user research                   Get some feedback
Evaluate competitors
Evaluate the current product
Watch people use the product

GATHER                                  EXPLORE

**CONVERGENT THINKING**

Find the patterns                       Revise and refine
Capture hunches                         Prioritize ideas and insights
Create a day-in-the-life narrative      Imagine building the site
Describe a task or scenario             Consider consequences
Visualize the problem

PROCESS                                 FOCUS

**Problem Statements**                  **Principles and Implications**
**Project Objectives**                  **Concepts and Big Ideas**
**Contextual Statements**               **Models**

# ACKNOWLEDGMENTS

This book has been percolating for a long time, the initial proposal dating back to the summer of 2014. I knew as I assembled it that it would be the most ambitious book I've written, dealing with a difficult, ill-defined aspect of our work. Jeffrey, Katel, and Jason gave me the chance of a lifetime to make it a reality. I'm grateful for their perseverance and faith. And I'm thankful to A Book Apart for giving practitioners like me a space to write practical guides about their work.

Lisa Maria Martin and I met at an IA Summit several years ago, when I rudely joined a lunch table where she sat with her colleagues. We had a great conversation and stayed in touch. Little did I know she'd be editing my book one day. Thanks to her, we cleared the cobwebs of confusion and clumsiness from the manuscript. Any that remains is due only to my stubbornness.

Several friends read early drafts of my proposal and offered insights, and their influence still lingers in these pages. I'm grateful to Erika Hall, Marc Rettig, and Russ Unger for giving me a lot to think about in those early days. Thanks also to Samantha Warren and Dan Mall, who kindly let me use examples of their work.

After I started writing, but long before I was done, I had the opportunity to teach workshops on discovery at three different conferences. The interest from participants buoyed my enthusiasm for the book, even during the toughest periods of writing. More than two hundred people attended, and I'm so grateful for the thought-provoking discussions.

It's perhaps fate that at the time I was working on this book I had several clients engage my firm EightShapes in discovery projects. I'm grateful to Brian Fox, Juliet Jones, Guillermo Ortiz de Zárate, Joe Pirret, Shawn Traylor, and Ray Whitney for their trust in me, for the opportunity to work on interesting projects, and for rewarding collaboration.

I've long been a fan of Kim Goodwin's work and writing. She's as busy as ever consulting, speaking, and traveling. I'm grateful she was able to make the time to craft a foreword that perfectly sets the tone for this book.

When I told Nathan, my business partner, that I wanted to write another book, he put on a good show, rolling his eyes and shaking his head. But he cheered me on when things got tough, challenged me to improve my thinking, and offered my guidance when I needed help. He remains the best partner a guy could have.

Nobody has taught me more about my writing process than Nicole Fenton. Where other editors helped me find my voice, Nicole helped me find out who I am, through writing. If that sounds challenging, you should know that she was transitioning jobs and embarking on motherhood all at the same time. Our work together changed me for the better, and I'm grateful for her enthusiastic efforts to make this book wonderful and to make me sound not quite so stodgy.

My kids were my escape, when I just couldn't spend another minute thinking about discovery. I'm grateful that they have just enough of my DNA to enjoy a board game or two at a moment's notice.

Most importantly, I thank my wife Sarah. She perhaps doesn't realize how much of an inspiration she was for this book. Discovery is, in a word, about learning, and Sarah's commitment to learning flows through every fiber of her being. Her natural capacity to gather, process, explore, and focus new information never ceases to amaze me, and she puts it to good use time and again, with every opportunity we encounter, every challenge we face.

# RESOURCES

Discovery relies on many different aspects of design and design thinking. These are but some of the resources I tapped to help me make sense of this process.

## The psychology, sociology, business, and history of creativity

- *The Myths of Innovation*, Scott Berkun
- *Creativity, Inc.: Overcoming the Unseen Forces That Stand in the Way of True Inspiration*, Ed Catmull and Amy Wallace
- *Creativity: The Psychology of Discovery and Invention*, Mihaly Csikszentmihalyi
- *Where Good Ideas Come From: The Natural History of Innovation*, Steven Johnson
- *Powers of Two: How Relationships Drive Creativity*, Joshua Wolf Shenk

## The practicalities of creativity

- *Syllabus: Notes from an Accidental Professor*, Lynda Barry
- *Gamestorming: A Playbook for Innovators, Rulebreakers, and Changemakers*, Dave Gray, Sunni Brown, and James Macanufo
- "Abductive Thinking and Sensemaking: The Drivers of Design Synthesis," Jon Kolko (http://bkaprt.com/pdd/06-01/)
- "Sensemaking and Framing: A Theoretical Reflection on Perspective in Design Synthesis," Jon Kolko (http://bkaprt.com/pdd/06-02/)
- Creative Whack Pack (deck of cards), United States Games Systems, Roger Von Oech (http://bkaprt.com/pdd/06-03/)

## Design methods and processes

- *Design Sprint: A Practical Guidebook for Building Great Digital Products*, Richard Banfield, C. Todd Lombardo, Trace Wax
- *Designing for the Digital Age: How to Create Human-Centered Products and Services*, Kim Goodwin

- *The Strategic Designer: Tools and Techniques for Managing the Design Process*, David Holston
- *UX for Lean Startups: Faster, Smarter User Experience Research and Design*, Laura Klein
- "The Redesign of the Design Process," Jared Spool (http://bkaprt.com/pdd/06-04/)

## Problem statements

- "Stop Doing What You're Told (Reframing the Design Problem)," Stephen Anderson as delivered at IA Summit 2013 (http://bkaprt.com/pdd/06-05/)
- "Defining the Problem: Q&A with Tom Chi," Luke Wroblewski (http://bkaprt.com/pdd/06-06/)
- "Defining the Problem: Q&A with Kevin Cheng," Luke Wroblewski (http://bkaprt.com/pdd/06-07/)

## Design principles

- Design Principles Guide, 18F (http://bkaprt.com/pdd/06-08/)
- "Creating Great Design Principles: Six Counter-Intuitive Tests," Jared Spool (http://bkaprt.com/pdd/06-09/)
- "Developing Design Principles," Luke Wroblewski (http://bkaprt.com/pdd/06-10/)
- Design Principles FTW (http://bkaprt.com/pdd/06-11/)

## Content

- *Nicely Said: Writing for the Web with Style and Purpose*, Nicole Fenton and Kate Kiefer Lee
- *The Elements of Content Strategy*, Erin Kissane
- *The Digital Crown: Winning at Content on the Web*, Ahava Leibtag

## User research and usability testing

- *Just Enough Research*, Erika Hall
- *Interviewing Users: How to Uncover Compelling Insights*, Steve Portigal

- *Handbook of Usability Testing: How to Plan, Design, and Conduct Effective Tests*, Jeffrey Rubin and Dana Chisnell
- *Convivial Toolbox: Generative Research for the Front End of Design*, Elizabeth Sanders and Pieter Stappers
- *The Essential Personal Lifecycle: Your Guide to Building and Using Personas*, Tamara Adlin and John Pruitt
- *The User is Always Right: A Practical Guide to Creating and Using Personas on the Web*, Steve Mulder and Ziv Yaar

### Sketching and visual thinking

- *The Doodle Revolution: Unlock the Power to Think Differently*, Sunni Brown
- *Back of the Napkin: Solving Problems and Selling Ideas with Pictures*, Dan Roam

### Psychological principles

- Mental Notes (deck of cards), Stephen P. Anderson (http://bkaprt.com/pdd/06-12/)
- *100 Things Every Designer Needs to Know About People*, Susan Weinschenk

### Documentation and creative briefs

- *Communicating Design: Developing Web Site Documentation for Design and Planning*, Dan Brown
- *Creating the Perfect Design Brief: How to Manage Design for Strategic Advantage*, Peter Phillips

### Conflict and collaboration

- *Designing Together: The Collaboration and Conflict Management Handbook for Creative Professionals*, Dan Brown
- Surviving Design Projects: The Card Game (deck of cards), Dan Brown (http://bkaprt.com/pdd/06-13/)

# REFERENCES

Shortened URLs are numbered sequentially; the related long URLs are listed below for reference.

## Chapter 1

01-01   http://www.designcouncil.org.uk/

## Chapter 2

02-01   http://gamestorming.com/
02-02   http://shop.oreilly.com/product/0636920021827.do
02-03   http://library.iasummit.org/podcasts/stop-doing-what-youre-told-reframing-the-design-problem/
02-04   bit.ly/badproblems
02-05   https://abookapart.com/products/just-enough-research
02-06   http://rosenfeldmedia.com/books/interviewing-users/
02-07   https://en.wikipedia.org/wiki/The_Humane_Interface
02-08   http://www.donnalichaw.com/the-users-journey/
02-09   http://sunnibrown.com/doodlerevolution/
02-10   http://www.danroam.com/the-back-of-the-napkin/

## Chapter 3

03-01   http://www.designprinciplesftw.com/collections/windows-user-experience-design-principles
03-02   http://designinginterfaces.com/
03-03   https://articles.uie.com/kj_technique/

## Chapter 5

05-01   http://optimalworkshop.com

## Resources

06-01   http://www.jonkolko.com/writingAbductiveThinking.php
06-02   http://www.jonkolko.com/writingSensemaking.php
06-03   https://creativewhack.com/product.php?productid=64

06-04  https://articles.uie.com/redesign_design_process/

06-05  http://library.iasummit.org/podcasts/stop-doing-what-youre-told-reframing-the-design-problem/

06-06  http://www.lukew.com/ff/entry.asp?336

06-07  http://www.lukew.com/ff/entry.asp?342

06-08  https://pages.18f.gov/design-principles-guide/

06-09  https://articles.uie.com/creating-design-principles/

06-10  http://www.lukew.com/ff/entry.asp?854

06-11  http://www.designprinciplesftw.com/

06-12  http://getmentalnotes.com/

06-13  https://www.thegamecrafter.com/games/surviving-design-projects-v2

# INDEX

## A

activities 12-15
affinity mapping 45
analysis paralysis 11, 32
Anderson, Stephen 23
assertions 17, 19
    direction-setting 57-58
assets review 39
Avatar 130

## B

Barry, Lynda 75
Berkun, Scott 54
big ideas 63-66
Brown, Sunni 19
bubble diagrams 53
budget 93

## C

Catmull, Ed 82, 92
Chalkmark 124
challenges 17
Chi, Tom 16
clickable prototype 24
Codeacademy.com 59
Code for America 58
concepts 63-66
consequences 88-89
constraints 17
content inventory 35
context
    business 26-29
    content 33-35
    user 31-33
contextual statements 17, 26-35

## D

day-in-the-life stories 48
deadlines 93
definition
    product 12
    terms 11
dependencies 87
design direction 120-121
documentation 117-125
    best practices 126
    discovery document 117
    objectives and outcomes 118
Double Diamond 14

## E

execution 6, 9
explore 13
exploring 89

## F

feedback 80-82
Five Whys 19
flowcharts 20, 71
focus 13
focusing 89
framework 17
Frankensteining 24

## G

gather 13
Google 16, 59
Google Docs 24, 47
Gothelf, Jeff 20
Gray, Dave 19

## T

## U

## V

## W

## ABOUT A BOOK APART

We cover the emerging and essential topics in web design and development with style, clarity, and above all, brevity—because working designer-developers can't afford to waste time.

## COLOPHON

The text is set in FF Yoga and its companion, FF Yoga Sans, both by Xavier Dupré. Headlines and cover are set in Titling Gothic by David Berlow.

This book was printed in the United States using FSC certivfied paper.

FSC
www.fsc.org

# ABOUT THE AUTHOR

**Dan Brown** is a web designer and cofounder of EightShapes, LLC, a Washington, DC-based design firm he opened with Nathan Curtis in 2006. Dan's portfolio includes work with Fortune 500 clients, nonprofits, industry associations, higher education, and the federal government. Prior to *Practical Design Discovery*, Dan wrote two books: *Communicating Design* (New Riders, 2011) explores how designers document their ideas and *Designing Together* (New Riders, 2013) promotes better collaboration. Teams all over the world have played his game Surviving Design Projects, to learn and practice conflict management skills. He tweets as @brownorama about design, management, board games, parenthood, and coffee.